BLUE ROOMS

ALSO BY JOHN JEROME

The Sports Illustrated *Book of Skiing*

The Death of the Automobile

On Mountains

Truck

The Sweet Spot in Time

Staying With It

Stone Work

Staying Supple

The Writing Trade

BLUE ROOMS

Ripples, Rivers, Pools, and Other Waters

John Jerome

henry holt and company
new york

Henry Holt and Company, Inc.
Publishers since 1866
115 West 18th Street
New York, New York 10011

Henry Holt® is a registered
trademark of Henry Holt and Company, Inc.

Published in Canada by Fitzhenry & Whiteside Ltd.,
195 Allstate Parkway, Markham, Ontario L3R 4T8.

Portions of this book appeared in different form in *Outside*,
the *Boston Globe Magazine*, *Adirondack Life*, and *New England Monthly*.

Library of Congress Cataloging-in-Publication Data
Jerome, John.
Blue rooms: ripples, rivers, pools, and other waters.—1st ed.
p. cm.
1. Rivers—United States. 2. Lakes—United States. I. Title.
GB1215.J47 1997 96-39325
910'.02169—dc21 CIP

ISBN 0-8050-2620-7

Henry Holt Books are available for special
promotions and premiums. For details contact:
Director, Special Markets.

First Edition—1997

Designed by Paula R. Szafranski

Printed in the United States of America
All first editions are printed on acid-free paper. ∞

1 3 5 7 9 10 8 6 4 2

For Mike and Jo, who got us started;
for Julie, who inherited the gene

Only from space can you see that our planet should not be called Earth, but rather Water, with specklike islands of dryness on which people, animals, and birds surprisingly find a place to live.

—Oleg Makarov, Soviet astronaut, *The Home Planet*

One would be acting in harmony with the laws of life, to which the phenomenal world is obliged to bow.

—Van Wyck Brooks, *The Flowering of New England*

contents

acknowledgments

This book would not have been possible without the advice, help, guidance, and support of Janet and Fritz Decker, Paul Jamieson, Bill MacLeish, Elizabeth Libbey, agents Georges Borchardt and Denise Shannon, editors Bill Strachan and Darcy Tromanhauser, and the staff of *Outside* magazine, particularly Laura Hohnhold and Mark Bryant. I am deeply grateful to all.

BLUE ROOMS

prologue: evening on rorschach pond

The late August day dawns breezy and bright, stirring an uneasy awareness: a summer is slipping through my fingers. I'm startled to realize that I haven't been on water or in it since mid-May. The days are getting shorter, the year itself is three-quarters gone, and I haven't yet accumulated the dosage of water and light I need to get through the winter to come. Not helping matters is the approach of my sixtieth birthday, implying a certain decrease in my personal supply of days. Sometimes I think Ishmael had it backward. Perhaps it wasn't the gray November of the soul that sent him to water, but a lack of contact with water that caused his misery.

According to the weather maps, a significant bubble of high pressure is sliding by to the southwest, promising fair weather. It'd be a nice time to go sit on a pond somewhere.

There's supposed to be an unspoiled small gem just over the New Hampshire border that my wife, Chris, and I have been wanting to check out for some time, but she's on a deadline and can't get away. I decide to go for an overnight look myself.

It's midafternoon by the time I have my gear together and the canoe on the car. Hubbard Pond is seventy miles away, an agreeable two-hour drive, and I pick my way on back roads through small towns, stopping to pick up a can of chili and some doughnuts (dinner, breakfast). The put-in point is hidden away on an unmarked dirt lane that takes half an hour to find, which I take as a good sign. Its tiny parking area is littered with the usual beer cans and broken glass—slobs' party trash, fishermen's garbage—but the pond itself is pretty enough. A bit weedy, surrounded by low green hills, it is maybe a mile across in its longest dimension. There are no cottages on it, in fact no shoreline interruptions at all except for a scout camp at one end, which I am hoping will be unoccupied. No boats in sight, no other cars around. It's a Tuesday; perhaps I'll have the place to myself.

I slip the canoe into the water, stow gear, and push off, enjoying again that eloquent small moment that always occurs when boat leaves shore. Pickerelweed and water lilies choke the fringes of the pond, dragging on the paddle, forcing me to choose my way slowly. It's a new canoe, unfamiliar, requiring a certain attention. I'd have preferred a bigger pond—paddle a longer distance, learn the boat's characteristics—but the farthest shore is only twenty minutes away. Okay, I'll change pace, or try to. This is an outing, not an expedition. I'll settle for a sunset and a dawn.

The whole place looks and sounds depressingly buggy, but except for a couple of pesky deerflies in midpond, I'm not significantly harassed. Dragonflies are everywhere, keeping the mosquito population in check. I hope. My information about the pond was unclear on whether camping was even possible, but at the far end I spot a well-worn site on a breezy point ten feet above water's edge. It's in a pine grove, giving a soft pine needle floor. There's a fire ring, and someone has even left a little firewood. It's quite nice. Behind the point is a marshy inlet, lit by late sun, from which the customary resident great blue heron rises ponderously as I paddle in. A red squirrel jabbers away overhead as I unload—irritated, I anthropomorphically assume, by my intrusion.

I pull the canoe up on shore, set up camp, lay out what I'll need for dinner, gather firewood to replace what I'll use, moving at a methodical, old-guy pace. Loose ends taken care of, I pour some bourbon into a cup of cool water, dig out the binoculars, and settle down with my back against a sun-warmed pine bole. At home I'd be watching the news about now. This is better. No need for a long paddle at all.

The binoculars, as usual, make everything ravishing. Staggered swaths of shining open water cut through clay-green lily pads, silvered trunks and upturned root stubs of fallen trees twist into outlandish shapes. Bumblebees work the vertical purple stalks of pickerelweed. Out of the corner of my eye I catch a red-tailed hawk hovering three feet above the marsh, but he disappears before I can get the glasses on him. The low sun bounces off the water, reflecting on the underside of the pine boughs above me, making a washed-out, artificial glow: faux light, coming from the wrong place. The ground beneath my butt, and the pine tree at my back, gradually become more comfortable. Oh, right, I remember, this is how the woods are.

I make a small fire, eat, clean up, put things away, keeping a careful rein on all physical processes—aware, as I would seldom be at home, that comfort and safety are easy enough to maintain but are in no one's hands but my own. A certain level of care is justified. Maybe that, too, is part of the appeal.

Dousing the fire, I go down to the shoreline to watch the sunset. Mount Monadnock—the actual mountain from which geology took the generic term—looms to the west, long shadows throwing their arms over its near shoulder. The wind and the pond grow quiet, and I think with pleasure of all the other ponds and lakes stretching beyond, the Adirondacks, Temagami, Boundary Waters, the ten thousand lakes, all coming under the same bell of high pressure, the weather fine, the sunset on its way, and all the other canoeists and campers settled down at water's edge, on an August evening, to watch the show.

The shift change is taking place. The sun reaches a certain angle and the diurnal species—insects, birds, fish, surely the vegetation too—begin handing the day over to the nocturnalists. The heat and the light change, and that drives everything else. For sensors as crude as mine it's not so much change as cessation, pause, stillness—and, usually, a soft glow from the sky.

Idly shifting the focus of the binoculars, I'm startled to turn up a spider, building her evening web on a fallen spruce twenty feet away. Through the binoculars the web looks big as a washtub, but when I lower the glasses it is ten inches in diameter and the spider is truly tiny. I saw her in the first place only because she's backlit.

She traces a spoke from the center down to the rim, moves over one space, traces the next spoke back to the center. She positively swoops from rim to center, in a rush, then goes off on another spoke, a shuttle in a circular loom. She switches modes and begins touching up the radials. She stops for a moment—out of breath, I suspect, spiders also needing to breathe—then darts off to some spot where the pattern is torn, and concocts a patch. She bounces the web as if to feel where its tension needs adjusting, dashing here and then there, adding strength, doing the engineering.

The backlighting from the descending sun fades and I can no longer see the web itself; I watch the spider work in empty air, conjecturing the pattern. Spiders, I'm told, have no extensor muscles in their legs. They have the customary flexors—for bending motions, for bringing food to mouth—but not the ones for reaching, for extending. For those motions they use hydraulic pressure instead, pumped up by the heart. They're little arachnid backhoes. If they become dehydrated, their blood pressure drops and they can't walk. For half of their movements they need a nervous system that flicks on a pump, rather than pulling on a muscle.

As I sit watching, the usual small voice in the back of my head begins wondering if there aren't more important things I should be thinking about. Nothing to do but eat, sleep, paddle, and look. And look and look. Nothing to look at but beauty. Detail. Sky. Light on water. In the woods perhaps one *can*, as the Zen master advises, just chop wood. Someone once came upon Daniel Boone, alone, a dozen miles from nowhere, lying on his back in a clearing in the woods, singing his heart out at the sky.

There's a waning three-quarter moon, and I tease myself with the idea of a moonlit paddle, fancying that I will somehow wake up and go do that, along about three A.M.; knowing better, I go out at dusk instead, grinning, startled once again that I am here: an old guy out paddling around in a pond just because it's such a pretty place to be. Circling over the mirrored surface, enjoying the feel of the paddle in the water, the gurgle at the stern, the pull in the shoulders. A beaver slaps his tail. He's thirty yards away, doing figure eights, once each lap expressing great distaste, with another slap and dive, at my presence.

Okay, it's your pond. With the drop in the breeze mosquitoes are coming out anyway, so I head for the tent. To the west the fading light organizes itself into stripes of orange in the sky and on the water as well, the far shore's silhouetted black treeline duplicating itself, a forest Rorschach laid down sideways. The moon over Monadnock is gorgeous, but to see it as sharply as I'd like requires the eyeglasses I've already stowed away for the night.

The object here seems to be to take the retinas to the woods. I crawl back into the tent musing on that. Get the retinas to water. Get them there, keep them comfortably alive, get them back out again. Solve the physical problems necessary to accomplish that. Enough. I lie back and, drifting off to sleep, review the retinal photo album of the day. Adventures in the phenomenal world: work out the physics and the rest is gravy.

Engram is the name science has reserved for a memory trace: the protoplasmic change to neural tissue left by a stimulus. If it exists, it will represent the physical location of a memory, explaining why memories last. Unfortunately we haven't found it yet: the engram remains inaccessible to observation. Experience, like memory itself, is written on water. Happily, water has a way of giving memory back. For some of us water is memory, and memory water.

1

the best river

> Twenty-three years since, when I was five years old, I was brought from Boston to this pond, away in the country. . . . That woodland vision for a long time made the drapery of my dreams. That sweet solitude my spirit seemed so early to require.
>
> —Thoreau's Journal, August 6, 1845

My family seemed to traffic in water stories long before I began to accumulate my own, repeating them, as families do, polishing them like prayer beads—as if once we get the stories right, if we ever can, we'll understand some additional thing. If we could only get into them what it really was about the incident, or the place, that made us notice it at the time, the story would tell us something new. Places that have meaning for us always seem to be telling us something, if not about the place then about ourselves. Water, I think, is always trying to tell us something.

One of my early memories is of sitting on a riverbank watching my father, Ralph, fish a dark pool. He stood just offshore in water to his knees, in dungarees and some kind of beat-up shoes—no fancy waders for him, in Oklahoma, during the Depression—and as he worked his casts methodically along the far shore, he told me about Indians and canoes, about how they could paddle silently, never lifting the blade out of the water. It was how they sneaked up on game, he said. Jud, my older brother, and I must someday learn to do this, so we could paddle for him while he fished.

The canoe was a running character in family tales, although I'm not sure I ever laid eyes on ours. It was probably pawned before I became conscious of such things. But it loomed large, for instance, in a family fishing trip to New Mexico before I was born. Ralph and Jud were fishing on a mountain lake, and Jud, not yet six, somehow contrived to drop a tackle box over the side. Ralph could see it resting on the bottom, looking within reach, but his arm's length didn't come close. Eventually he stretched as far as he could with the tip of his fly rod and still couldn't touch it. The point of the story, repeated over the years, was how clear the water was. It is as if all three of them—Ralph, Jud, and my mother, Gwen—were determined to teach me that in water, clarity is everything.

I always thought I'd discarded those legends. Now I see that if my elders told me something was worthwhile, I was virtually fated to spend the rest of my life trying to check it out. They oversold a good many things, but clear water was not one of them.

The riverbank where I sat was the Mountain Fork, in the Kiamichi Mountains, a peculiar little east-west range tucked away in the southeast corner of Oklahoma—foothills of the Ouchitas, which in turn are foothills of the Ozarks. Few people outside the state have ever heard of the Kiamichis, but my parents loved the place, and it was the first place-name I remember.

In a prairie state one goes to mountains for relief, which in our case was not exclusively topographical. Ralph was an alcoholic and a fisherman, in that order, and the Mountain Fork was his favorite river, in part because it slipped across the Arkansas border and back again, and in those days Oklahoma was dry and Arkansas was wet. This sometimes made the Kiamichis less than restorative for Ralph, however refreshing they may have been for the rest of us.

According to Jud it was usually Gwen who got us there, fleeing Oklahoma City heat in the summer night, canoe and fly rods lashed to the roof of the car, kids and semiconscious husband sprawled backseat and front. I don't remember any of those trips, but by age four I could rattle off in order the last few towns—Hugo, Idabel, Broken Bow, Bethel—before we reached the river. Our destination was always one of the small cabins owned by a half-breed Choctaw named Coleman Ward.

So the Mountain Fork was my first river. All I really knew of it was the stretch directly across the dusty road from the Wards' place, where a ledge made a waist-high waterfall, with a quiet pool above and a foamy, slick-rock playground below. I remember floating above Ralph's naked back, his ill-shaven chin scratchy on my arms around his neck, as he breaststroked out to an exposed rock in the middle of the pool above the falls. I assumed everyone's father did that; decades later I would swim with my own small children the same way, and the memory always came back. For Jud and me that river represented paradise, purest freedom. Whatever pleasure I take in the natural world began for me there.

When Jud became ill, at sixty-four, I began thinking of

the Kiamichis again, wondering what that country was really like. I hadn't laid eyes on it since I was eleven years old, and over the intervening years it had become almost mythological for me. I ordered topographical maps of the area and would stare at the wriggling brown lines, attempting to rebuild the landscape in memory, imagining what could have happened to it. In our last conversations Jud and I were comparing fifty-year-old memories of the place; when he died I decided to go take a look.

I went in late September, hoping to escape the more oppressive summer heat, approaching this time not from Oklahoma City but from Texarkana to the south, the nearest commercial airport. That corner of Arkansas is flat, closed in: scrub oak and big pines, red-dirt gumbo roads torn by logging trucks, stinking paper mills. The forest alongside the highway would occasionally open into scrubby pasture land where mottled grayish Brangus cattle gathered at the edges, seeking shade. Brilliant white cattle egrets gathered with them. (Cattle egrets—I'd forgotten about them. That would become an interior refrain while I was down there: Gee, I'd forgotten that, forgotten all that.) Small fundamentalist churches were everywhere, four or five of them to every gas station or roadside store. One church had a fireworks booth in its front yard. The crossroad towns were mostly boarded up, only video rental outlets and pawn shops still viable. The afternoon temperature was in the mid-nineties.

Just after crossing the Oklahoma border I spotted Weyerhauser's Kiamichi Tree Farm—the first occurrence of the name that I'd come to regard as a secret password. Then another sign pointed to Mountain Fork State Park: the river I'd come to see, although I was still thirty miles too far south. The dirt road into the park transected an extensive clear-cut, now replanted with pine seedlings two feet high. The park, in an oak grove, consisted of a strip of gravel, a few parking spaces with charcoal grills, and restrooms. One camper pickup was in residence, owner gone fishing. The river, blue-green but milkier than I remembered, was blessedly cool to my hand. Upstream of the camping area, behind a good-size concrete dam, lay a drowned forest, one of those awful-looking backwater messes that have become something of a standard feature in the Southwest.

Back on the pavement I turned left in the town of Broken Bow and headed north. My topo map had shown another reservoir, a huge one, now backing up the Mountain Fork to within a few miles of Coleman's place, but his establishment was still on the map, a collection of small dots. As I headed for those dots, side roads darted off to marinas, bait shops, a golf course, but I ignored them, knowing where paradise lay. It would be just above a stretch of the river called The Narrows, where Boktuklo Creek comes into the Mountain Fork from the west.

As a child, my two main heroes were Tarzan and Robin Hood. Out of the entire pantheon of children's literature I chose the two who got to live in the woods. The Kiamichi Mountains were responsible. When I was sixteen I read *For Whom the Bell Tolls* and was swept away from the opening scene, with Robert Jordan lying on the floor of a pine forest, scouting the enemy. When I read it I could smell the pine needles under his chin, but I wasn't in the Pyrenees, I was in the Kiamichis.

How our family first connected with the Wards is now lost to memory. Their main cabin had once been a store and filling station—gas pumps still stood outside—but had become the family residence by the time I knew the place. The shelves on its walls were lined with jars of home-canned vegetables and of venison, a staple in the Ward diet. There was a small garden, a low barn and barn lot, chickens and pigs. Jud remembered following Coleman's horse-drawn plow, picking potatoes from the freshly turned furrow. They called their establishment a "tourist court," but it was really only a subsistence farm with a couple of extra cabins to rent to fishermen.

Coleman was about sixty when I knew him, a small man with stark white hair and weathered face, fiercely proud of his Choctaw heritage—"Injun blood," as he said it. He affected a single white feather twisted into his hair, and taught me how to do the same, making me a kind of honorary Indian. He had a large extended family which he ruled absolutely, handing down prickly Indian wisdom, the patriarch as amiable contrarian. He poked fun at everyone but particularly at white men, whom he always portrayed as incomprehensibly but hilariously stupid. I had no living grandfathers and elected him to the role, following him around like a puppy. Jud remembered him as a bit of a clown—which he was, trickster version. His Choctaw name, Gray Fox, was perfectly symbolic.

I exaggerate my familiarity with the Mountain Fork even to myself. I was probably there fewer than half a dozen times before Ralph and Gwen were divorced, and returned only once after that, on a camping trip with Jud's Boy Scout patrol, chaperoned by Ralph and our new stepmother, Evelyn. Ralph served as informal scoutmaster at first, helping the boys set up camp, but after a day or two he somehow sliced open his hand and withdrew to the cabin, not to reappear, essentially, until time to leave. Jud and I were relieved. The scouts immediately pulled up stakes and moved to a new campsite on the other side of the river, effectively beyond adult supervision. It was summer camp run by the kids, Brigadoon with ticks and poison ivy.

I had the time of my life. I was only occasionally permitted to stay overnight with the older boys, but I swam across most mornings to join them, to play all day on the riverbank rope swing that they'd put up. When I wasn't with the scouts I played with Coleman's grandson, called Teacup, who was about my age. When we weren't swimming we roamed the woods behind the cabins, usually playing Indians against imaginary cowboys. (I refused to play the white man.) Teacup's younger sister Felicia—"Witchie"—joined in when we'd let her, a natural in the role of dark-eyed Indian maiden. I was extremely conscious that I was playing with Indians, of whom I was a little frightened. They were tougher, faster, stronger than the kids I knew, putting the river and the surrounding hills and woods to more thorough use than I'd ever imagined possible. It was a peculiar early introduction to a kind of benevolent racism. They knew from their grandfather that Indians were superior to white men, and there I was demonstrating to them, as they demonstrated to me, that he was correct. I didn't mind a bit; I'd have preferred being Indian myself.

Driving up U.S. 259 nearly five decades later, I spotted the Coleman Ward place immediately, a mile or so upstream of the Boktuklo bridge. The one remaining cabin, boarded up, was now flanked by two neat, modern, middle-class houses. A small sign marked the red-dirt road up the hollow where Teacup and I had killed so many cowboys: JOE WARD'S POULTRY FARM. Good, it was still in the family.

There was no sign of life, though, so I pulled up across the road and made my way down to the river, to stare in horror at a sun-blasted, sluggish, soapy gray sump. A low-water concrete bridge now crossed a few feet downstream of the ledge that once made my perfect swimming hole. The smell was rank.

I walked across the bridge and climbed the opposite hill into what should have been deep natural forest but was now perfect rows of uniform pines exactly eight feet tall. I turned around and walked back across the bridge, looking for the rock we used to swim to, but couldn't find it. The heat was suffocating; after ten minutes out of the rental car I dodged back in and cranked the air conditioner to high. It occurred to me that I might not be tough enough for this country anymore. There used to be wild places that were paradises; soon, I found myself thinking, the only wild places left will be the hells.

Roadside graffito: TAKE A TICK BACK TO TEXAS. When I'd come there as a kid I ran free most of the day, interfered with by adults only at meals and bedtime. Each night, though, I had to undergo a strip search for ticks. I would stand naked, squirming, beside a table with a kerosene lantern on it, while Evelyn methodically inspected every inch and cranny of my hairless eleven-year-old hide. When she found one she would pluck it gently and scrape it from her fingernail into the chimney of the lamp, where it would sizzle and make a tiny flare in the dark cabin. It was humiliating but also titillating. I didn't really mind that, either.

I stopped at the chicken farm to inquire. I didn't remember Joe, who was Teacup's older brother, and he didn't remember me or Jud but did recall Ralph and Evelyn. He was reticent, shy, not terribly interested in the state of the river fifty years ago. He told me that Teacup—now called Leon—taught college in Denison, Texas, and Felicia Ward Ryan, a former dancer and widow of a war hero, lived in Norman.

I got back in the car, found a shady place under the Boktuklo bridge, and stopped to think. Someone had strung a trotline across the Mountain Fork just upstream of the creek's mouth. Mosquitoes whined. Another low-water bridge crossed the river here, but had fallen in and was impassable. A dome of heat was hammering the Southwest, making a peculiar glaring haze. It was so hot I kept thinking there must be some mistake. I was unable to overcome my reluctance to get out of the car, thinking of ticks and chiggers. The rattlesnake scene in *True Grit* was set just a few mountains to the north. When I'd come here with the Boy Scouts, one of them got poison sumac in his mouth, of all places. In the low-lying, swampy, southeast corner of the county there are actually alligators.

Better first to get the lay of the land. I headed back toward Broken Bow, home of the only combination pawn shop–hat store I'd ever seen, for some catfish and an air-conditioned bed.

• • •

Coleman's standard fisherman's lunch: sprinkle a little salt onto some wax paper, fold it, put it in your shirt pocket. When you get hungry, splash some minnows onto the bank, dry them on a rock in the sun, salt and eat them like canapés. That's what he said, but I never saw him do it. I never succeeded in splashing up a single minnow. I doubt I'd have eaten it if I had.

Near the end of the scouts' stay, Coleman organized a daylong group hunt. His four adult sons joined in, plus some other male relatives and Teacup. Jud and his scout patrol were invited, and I talked my way into tagging along; Ralph stayed in the cabin. We started the morning "noodling"—wading nose-deep in a nearby slough to grapple fish by hand from under logs and mud banks. I was advised that there were snakes and snapping turtles as well as sharp-finned catfish under there—which, as intended, turned it into the most frightening act I've ever performed, before or since. The men would occasionally come up with some wildly flapping thing and stuff it into a gunny sack, with great whooping and laughter, teasing the kids about our failure to catch anything. Live things occasionally brushed my fingertips or ankles, but I never seriously considered attempting to grab anything.

The group then divided up, going off in different directions in pursuit of any other edible protein the hills might provide. One .22 rifle and one bareback horse were available; someone took both and wandered off to hunt deer, just as illegal in the off-season as the noodling had been. One of Coleman's participating sons happened to be the game warden. How were you going to live if you couldn't take game?

I lost track of Teacup—my assignment was to stay with Jud and keep out of trouble—and didn't see him again until afternoon. He'd gone off with an uncle who wanted to fish a stock pond with rod and reel. The uncle hooked a good-size bass, but it took his plug. Unaccountably, at least to me now, the uncle sent Teacup in to grapple the fish to land. Just about the time we walked up to the hole, Teacup stood up in knee-deep water, dripping wet, with a three-pound large-mouth bass flapping from his forearm, one hook in the fish's mouth, another embedded in his wrist. Somebody got the fish off the hook. Teacup listened impassively—I was staring in horror—while the adults argued over whether or not the hook had snagged a tendon. If not, it could simply be cut out. If a tendon was involved, the hook would have to be pushed on through the flesh to emerge at a second point; then the barb could be clipped off and the hook backed out. Finally they decided to let the women solve the problem and sent Teacup home on the horse. My last sight of him was riding off toward the cabin, the bass plug still dangling from his arm.

Breakfast in Broken Bow, early: two young construction workers try, loudly, to sober up sufficiently to go to work. The other customers pointedly ignore them. An older man in a gimme cap sits alone, drinking coffee and smoking cigarettes, addressing the patrons as they enter and leave. He is a fixture, I realize, almost a tradition in a breakfast joint like this.

"You see my pickup?" he asks one arriving customer.

"Yeah, looks pretty good."

"Well, you seen the wrong side of it then."

• • •

You can drive completely around the Kiamichis in a day, which I set out to do, wanting a better grasp of the watershed. The mountains have an Allegheny feel to them, albeit on an east-west bias: long parallel ridges with deep valleys between, geologic corduroy. A series of south-flowing rivers cut through them, starting with the Kiamichi River on the west, then the Glover, the Little, and the Mountain Fork. The Kiamichi runs into the Red River (quavering harmonica), which forms the border between Texas and Oklahoma; the Glover and the Mountain Fork are collected by the Little, which swings east and wanders into Arkansas before connecting with the Red. The Glover is alleged to be the last free-running river in Oklahoma, all the others having been dammed to make reservoirs. In a region where even muddy stock tanks draw recreational skinny-dippers, decades of furious dam building have left Oklahoma claiming more surface water than Minnesota.

On my drive around the Kiamichis I visited a succession of these dams. Each had a ranger's office, some RV camping spaces, and a great hot, steaming lake stretching off into the distance. Stumps usually dotted the surface of lukewarm milky brown water. Children's footprints scuffed the little stretches of weedy sand that served as swimming beaches. No shade, no relief. Out on the water I would see an occasional bass boat, but it looked more like punishment out there than recreation.

My circuit of the mountains fetched up in Mena, Arkansas, population 5,154, where I sought the headwaters of the Mountain Fork—properly, the Mountain Fork of the Little River. The local paper was abuzz over the mysterious seizure by treasury agents of a private aircraft at the town's small airport—although the two Colombians who flew it there had been released. Town fathers seemed to be worried about getting a reputation as an international drug exchange point—more worried about that, anyway, than about the boilingly nasty conservation arguments that filled the letters-to-the-editors columns. There was a rack of fringe-group hate literature in the motel lobby, including one pamphlet aimed at environmentalists, ". . . whose real aim is not a cleaner environment, but a *socialist* environment." Arkansas' license plate motto: "The Natural State."

In the restaurant that night the two couples at the next table called for their check. "Sure enough," said the waitress, stubbing out her cigarette. "J'all get *full*?"

When I flipped on the TV at 6:30 A.M., before heading out to look for those headwaters, a nature program was explaining the complex damage that clear-cutting does to watersheds.

On this Saturday morning a good portion of the male population of Arkansas seemed to be gathered at filling stations, examining the insides of tire casings. Local dogs were out in force, touring the back roads in troupes of three or four. The Mountain Fork starts in Oklahoma as a small brook but crosses into Arkansas to pick up substantial water from various feeder creeks before heading back into Oklahoma. I picked my way on a succession of gravel roads, stopping at every bridge, never quite sure which creek I was on, looking for something identifiable as the Mountain Fork. Eventually enough small feeders came together to make a small river, and consolidate their upland valley in the process.

Most of the time it was a woodsy little stream with the

usual riffle-pool-riffle progression; a long pool would turn the water olive green, then a riffle would clear it up again. The road cut away from the river above Hatfield, Arkansas, just before a long series of rocky flats where I'd been told there was fine whitewater sport in spring's high water. As soon as I turned back toward Hatfield I was into more clear-cutting and lost the river, unable to pick it up again—by car—until I was back over the Oklahoma border. Once there I stopped in Smithville, four or five miles upstream of the Wards' homestead. When I inquired about Coleman at Jess Umstead's store, faces lit up. The town seemed to remember him fondly.

How come, I asked (slipping into the vernacular), they messed up that beautiful stretch of river down by Coleman's place? Well, that was the Highway Department, said the woman at the counter. When they built the road they cut down the big walnuts that used to make shade. And yes, the river was unpleasant along there now, wasn't it? I paid for my Coke and turned to leave. A huge man, a Jim Thorpe look-alike who had been listening quietly to our conversation, finally spoke: "Weyerhauser," he said. Two-beat pause. "And beaver."

U.S. 259, the major north-south route in eastern Oklahoma, wasn't completed until 1962. Coleman and his wife Willy owned frontage on the dirt road it supplanted but didn't own the riverbank. When the new highway came through, the state not only cut down the walnuts but bulldozed the banks for road fill. There was nothing the Wards could do. Weyerhauser then bought up the surrounding hills, built low-water bridges for access, and began clear-cutting. (Weyerhauser acknowledges the "potentially disastrous" effect of clear-cutting on major streams, through erosion and stream siltation, but claims to have adopted stream-side management regulations "to ensure that silviculture practices are compatible with the preservation of the river.") So the fine new road that provided all that handy access up and down the eastern part of the state, that now brought the tourists to Broken Bow Lake, had wiped out Coleman's little kingdom. It was the highway, or Progress, or the greatest good for the greatest number, that had so damaged the river.

I found Smithville's small cemetery: Coleman Ward, 1883–1961; Willy, 1886–1972. Maybe he died before the river was ruined. The cemetery was overgrown, and my shoelaces became clotted with cockleburs. Back in the car, when I tried to remove these pernicious little devils, my fingertips were soon bleeding all over my sneakers. Somehow it seemed a perfect Coleman Ward joke.

From the Oklahoma Scenic Rivers Commission report on the Upper Mountain Fork: "Numerous logging low water bridges cross the stream. These allow the canoeists to experience the view of the logging operation, the truckloads of timber being carried away by the heavy vehicles." Boy, *that* ought to bring out the nature lovers.

I headed downstream to Broken Bow Lake, in my prejudiced view the true impediment to the unvexed progress of the Mountain Fork. Its enormous dam was completed in 1969, for the usual rationales of flood control, recreation, hydroelectric power, water supply, and fish and wildlife management. These mountains get fifty inches of rainfall a year, and the need for flood control is genuine. A brochure claimed the lake had 180 miles of shoreline stretching 22 miles back upstream. Huge. It was prettier than some of the other reservoirs I'd visited—no protruding stumps, and the water actually looked blue instead of brown—but I didn't like it much. Its most unappealing aspect was the dead zone around its edges, the result of raising and lowering the waterline—an acid yellow strip of death around every island and every shoreline, as if some Corps of Engineers Christo had been arrogantly demonstrating his power to change the landscape.

But I was just in a snit because a few miles of childhood memories had been made ugly. Easy come, easy go: it couldn't have mattered much to anyone but the Wards, and my family and me, and the few others—a few hundred, a few thousand—who knew the place, who still remembered it. The entire Mountain Fork is only sixty miles long. Reservoirs are more democratic recreation, and at least better for the surrounding countryside than the institutionalized mayhem of clear-cutting.

Just below the dam was Beavers Bend State Park, a neat little facility built in 1937 by CCC crews, which contained a mile or two of carefully maintained river. Rental cottages overlooked a manicured swimming hole—replacing a spot downstream, a ranger told me, that had been beloved of locals for decades but now was subject to dangerous surges from the dam. Canoes and paddleboats could be rented; there was a riding stable and the usual nature trails, and plenty of access to the regularly restocked trout.* A self-serving small

*Trout were unheard of in Oklahoma in my father's day, any natural water being far too warm for their survival. The depth of the reservoir, however—over eighty feet—keeps the water cool enough to allow hatchery fish to stay alive, at least long enough for anglers to haul them out again.

museum celebrated the lumber industry's contribution to the local economy. (Farther north another state park commemorates the Choctaw village of Hochatown, drowned by the reservoir.)

I rented a canoe and signed up for a group pickup at the end of a ten-mile run below the dam. Two couples had also signed up for the float. At the put-in we waited while a state fish hatchery truck dumped in a load of trout, then slid three canoes into the water. The river smell that was too strong back at Coleman's was just right here—the river I remembered. I pulled over into a cypress grove to let the others go, wanting the river to myself. While I waited—still sweating from helping unload canoes—I slipped over the side into delicious water. It was eleven in the morning, I had a picnic lunch and no other responsibility than to be at the pickup point at five. Once I was wet, the heat was no longer oppressive. In fact it was a gorgeous, hot, sunny day on a cool, clean river just like I remembered. My childhood memory was not so far off base. Maybe, I thought, they didn't destroy my river, they just moved it thirty miles downstream.

Kingfishers dipped and darted ahead as I paddled; cicadas screamed. The river was fine canoeing, muscular in its green upwellings, a tumbling, caressing delight. Between riffles I gawked at huge cypresses, of which I had no memory. (Their northernmost range mysteriously stops just below Boktuklo Creek.) Whenever I no longer felt air-conditioned, I slid into the water. Once I stopped to photograph some bankside Indian paintbrush and let the canoe get away from me, and had to chase it, giggling, across a gravel bank, knee-deep, thigh-deep, holding my camera high out of the splash. At moderately spaced intervals, for a few minutes at a time, I was eleven years old again, the river as delightful and the world as wild, endearing, and rich with possibilities as it had ever been.

The first few miles were a succession of pools separated by small ledges, natural dams just like the one that I remembered at Coleman's place. When the river flattened out I caught up with the other two canoes for lunch, and thereafter it was a slow, flat-water pull through agricultural land: cabins along the banks, power lines, noisy irrigation pumps, wading cattle. (Access by cattle messes up a river instantly, a phenomenon one sees as clearly in Vermont as in Oklahoma.) We passed one saloon, where a bridge crossed the river. We were never out of earshot of highways.

On the way home the outfitter who picked us up told me that the Mountain Fork offers the best smallmouth bass fishing in the state. People catch a hundred fish a day. They don't eat them, though. Over near Mena—in "The Natural State"—there'd been a treatment pit where they soaked the logs in poison to keep rot (i.e., life) from getting into the lumber. It was designated a Superfund site and allegedly cleaned up, but insufficiently. Chemicals, the driver told us, still seep into the Mountain Fork. Such things had never crossed my mind when I'd been swimming.

I'd been misled by the memory of primitive cabins: despite what befell the river, the Wards had prospered. They were more substantial citizens than I'd realized. Teacup remembered me well. His full name is Judson Leon Tecumseh Ward, I was surprised to learn, after my brother; his mother had stayed with Gwen and Ralph in Oklahoma City during the last stages of her pregnancy. He's now a professor of literature, a tournament bridge player.

I asked him about the Choctaw blood. Coleman's old joke was that since he and Willy were both half-breeds, together they added up to one whole Indian; state records, however, listed him as one-thirty-second Choctaw. "That was only to get out from under the government," Leon said. If you were half blood or more, the government retained a guardianship role, so people listed themselves with as little blood as possible. Coleman and Willy were in fact half bloods, and happened to be fourth cousins: their grandfathers had been brothers, white men who came to the territory to teach in Indian schools, and took Indian wives. Coleman spoke fluent Choctaw—"sometimes he would barely speak anything else," said Leon—but he wrote newspaper columns, in English, for the *Daily Oklahoman* and the New Orleans *Picayune*. He was a voracious reader and, Leon surprised me by saying, "something of a twentieth-century Thoreau transcendentalist." Maybe that's where the contrariness came from.

The incident with the bass plug had been less dramatic, perhaps, than I remembered. Not a severe wound: Leon doesn't remember the removal of the hook, and has no scars from it. But he was a small kid at the time; how could his uncle have sent him in to grapple the bass? "Well," said Leon, "you have to realize, we were practically amphibious." That seems absolutely right. We spent most of our time in the river. Just to get to the scouts' camp you had to swim, an inconvenience that didn't seem to bother anyone. The scouts may have used some kind of boat for purposes of supply, but I don't remember it. I do remember sidestroking across and back daily, holding my clothes overhead in my free hand to keep them dry.

Shortly before he died, Jud reminded me of the Zen advice about not pushing the river. Perhaps that's what I am doing with the Mountain Fork. My father was a tale-teller, my brother a poet, and I surely inherited the habit of exaggeration. The Kiamichis were perhaps never as idyllic as I'd remembered, and the river probably not as devastated as I perceived it now; perhaps Coleman's old stretch was simply a bad segment. Surely it was I who had changed, more than the river, losing the child's ability to hunker down and explore every nuance of a stream, every pothole and rock surface, with bare feet and hands. Maybe it's just that when you're eleven years old, any woods, any water, is paradise. It was the best river because my father said it was. I had my grievances with him, but I didn't go there to find out that he was a fool. It was the best river at the time, the best he or any of the rest of us knew. I've been enjoying it ever since. There's no reason to stop that now, even if the river he and I remembered no longer exists.

2

pools and riffles

> It seemed the only right way to enter this country, borne on the bosom of the flood which receives the tribute of its innumerable vales. The river was the only key adequate to unlock its maze.
>
> —THOREAU'S JOURNAL, UNDATED

Water draws the only straight lines in nature. It is in its curvatures, however—in movement—that its true character comes most clear: where a velvety layer of limpid water spills over granite ledge, for instance, defining liquidity. This should not surprise us. The earth is beautiful because of water.

In middle age Chris and I were surprised to find ourselves developing a joint interest in rivers. This came about, awkwardly enough, for reasons more spirituous than spiritual. Friends with a canoe invited us along one summer afternoon to float down a stretch of the nearby Deerfield. We had a fine time, and went again—several times—until we'd almost ritualized the experience. We would take along a picnic and a cooler of grapefruit juice laced with gin, and spend the day floating and swimming and getting boozily sunburned, never intending more than a wet afternoon. It beat croquet all hollow. It was only a fancy way of going tubing, really, and in fact to reduce the crowding in the canoe we started bringing along cheap inflatables. The current was slow, however, and paddling the inflatables was tiring, towing them behind the canoe only a little less so. We ended up buying an inexpensive canoe of our own.

The Deerfield is a charming small New England river, rising in the Green Mountains of Vermont, flowing south across the Massachusetts border and then east into the much larger Connecticut River. It maintains its appeal despite being dammed into two large reservoirs and swinging by to cool the reactors of Yankee Rowe, the oldest nuclear power plant in the country (now deactivated). There's heavily trafficked whitewater in the Deerfield's midreaches, but our interest was only in the last twenty miles or so, which offer a selection of totally undemanding float trips in reasonably clean water. Every few miles there's a rope swing of some kind, giving aerial access to a deep hole. Chris loved the bird life, I loved the water that attracted it; with our own canoe we began quietly going off to float by ourselves, sometimes even without the gin. The banks of the lower river are muddy, and at the takeout the smell always reminded me of Ralph.

Lakes are hard to fall in love with, too large to embrace, pieces of physical geography chiefly useful for providing the best possible setting for sunsets and dawns. To love a lake is to love its situation, its individual configuration of horizon and sky. (The same applies to oceans: "I don't care for the sea," Adirondack writer Paul Jamieson once confessed in a letter to Chris. "It is just an excess of water making a great commotion without getting anywhere. Give me a river that runs, runs fast or slow, straight or meandering, through ever changing vistas of the fruitful land.")

Nobody, however, doesn't love rivers. Driving along the Deerfield one late afternoon of a raw early-April day, I was amazed to see cars parked beside the river, rowboats out on it, the smoke from charcoal braziers along shore. The water was low, exposing mudflat banks, and what little vegetation there was was uniformly gray and tattered. The temperature was a chilly forty degrees with a nasty little north breeze. But people were out on the water, and alongside it, recreating.

Where a bridge crosses a river there will be four paths leading down to the water: on each side of the road, at each end of the bridge. There's always a path alongside a river, or for that matter any body of water, anywhere. You can be set down by float plane in the northern wilds, five hundred miles from anywhere, and there will be a shoreline path; the wildlife (or what hunters call "game") will have made one if human beings have not.

How gratified we always are, in a car, when the road follows alongside a river. How irritated we always are, in a canoe, when the river we're on has a goddamned road following along its banks.

Because the Deerfield is used for the generation of electricity, it is drawn down and allowed to rise again according to periodic needs. Warnings of sudden rises are posted along the shore. Whitewater outfitters coordinate with the power station operators, but we scheduled our float trips on a catch-as-catch-can basis, so ineffectually that we began referring to them jokingly as wet-walks, for the amount of time we spent wading, dragging our watercraft over exposed shallows. We usually picnicked on one island or another, sometimes on a shingle that would gradually disappear from under our feet as an upstream release raised the water level.

The Deerfield is very gentle, but even in the clumsy inflatables we were charmed and intrigued by the whorls and shifts of current around the bends, the curlicues, the physics of the flow. Thrashing our way down the occasional riffles, capsizing more than once—with great hilarity—we learned almost nothing about how a river works. Whitewater enthusiasts talk about reading the river, but we hardly noticed there was a text. All we were doing was floating down our local stream.

That was our mistake: rivers' natural restlessness makes them endlessly complex and mysterious, adding to their already considerable entertainment value. Water's search for its own level, for example, is so imprinted on our brains that we have trouble comprehending the capricious pitch of river surfaces. Downhill is not a river's only slant. It bulges in the middle, for instance. Flowing water is slowed by friction with the bottom and the banks; the drag pushes water toward the center of the stream, actually making the surface of the stream slightly convex. The tendency of surface water to move toward the center of the flow causes water-soaked logs and other debris to gather there, even though it is in effect the top of a hill. Houseboaters on the Mississippi sometimes moor to these logs to hold their craft in midstream.

Surface tension also slows the flow, so the zone of fastest water—what zealous canoeists refer to as "the river within the river"—is a little below the surface and equidistant from the banks. (Heat from the friction of this flow is what causes frozen streams in spring to melt from the center outward.) Or rather the river within the river stays in midstream until the river curves; then the faster flow swings toward the outer bank, where its momentum stacks it higher than the rest of the surface, banking the surface of the water in the curve like a properly engineered road.

In a topographical sense, a river's water is not as important as what it carries with it. Rivers are sometimes better understood not as conduits but as carving machines. "There is a rough physical law," says Wallace Stegner in *Beyond the Hundredth Meridian*, "to the effect that the carrying power of water increases as the sixth power of its velocity, which is to say that a stream moving two miles an hour will carry particles sixty-four times as large as the same stream moving one mile an hour, and that one moving ten miles an hour will carry particles a million times as great. A stream that in low water will deposit even its fine silt and sand, in high water will roll enormous boulders along its bed, and sometimes one can stand near the bank and see a rock that looks as big as a small house yield and sway with the force of the current."

So streams not only carry particles but sort them by size, dropping out the larger as the current begins to slow, carrying the finer material onward. When a river curves, water on the outside of the curve must move faster than the water on the inside of the curve (just as on a curved roadway, the outside automobile wheel, covering more distance, rotates faster than the inside wheel, which is why automobiles must have differential gears between their driving wheels). Slow water on the inside drops its load, creating a sand bar; fast water on the outside cuts away at the outer, concave bank, deepening its curvature—and picking up more material in the bargain. The curve grows.

Rivers also will not run straight (unless the Army Corps of Engineers gets its hands on them). Fast water eating at the outside bank deepens the channel below it. The drag of the bank quickly slows that water, however, making it drop its load downstream along the same bank, building a bar that eventually forces the current to swing back across the channel. When that current hits the opposite bank it begins again to cut it away, re-creating the dynamics above. The tendency of rivers is thus to curve more rather than less, which may be why the Corps of Engineers' traditional straight-line approach is so naturally offensive to the eye. "The line of beauty," said Thoreau, "is a curve."

The line connecting the deepest parts of the channel—the middle line of a navigable channel—is called the *thalweg*, from *thal*, valley, plus *weg*, way. (It may be blithely disregarded in a canoe: with only inches of draft, water depth is seldom a critical consideration.) Like the current, the thalweg migrates back and forth across the bottom, even in the rare straight channel. Where the current cuts into a bank it begins to carve a *meander*, named after what is now the Menderes River in Turkey. The sinuosities of such channels are the same, no matter the size of the river: without an index of scale you can't tell whether you're looking at the map of a rivulet or a great river.

Looking for other places to paddle close to home, Chris and I tried the Deerfield's reservoirs. Both are surrounded by power company land, both are encircled at low water by the standard hydroelectric dead zone, the river bed at their outlets sometimes going completely dry between power generations. Harriman Reservoir, a few miles over the Vermont border, was nice enough in midweek but on weekends filled with water skiers and those obnoxious jet skis, the watersport equivalent of snowmobiles. Somerset Reservoir, farther north and at the end of ten miles of dirt road, was too remote for the high-speed crowd—we were startled to see a loon there, a tribute to the lack of traffic—but permitted no overnight camping, and by then we were beginning to want more than an occasional afternoon on the water.*

Our appetite for rivers then took us, by a roundabout route, to the Adirondacks' Raquette. It is the most heavily used recreational river in that park, but we discovered that in the spring we could have one stage of it—north of the town of Long Locke—virtually to ourselves for two- or three-day trips. With luck we would slip through after ice-out but before the Adirondacks' notorious blackflies made outdoor activity miserable. When we did we were shown the sublimities of a highland river at flood stage—a prospect and an experience so quietly magnificent that we've since made it a point to go back nearly every May. Except for one drop of eighty feet in a mile—at Raquette Falls, the sole portage on this stage—it is in springtime a broad, flat, dark, slow-moving conveyor belt meandering through drowned flatlands, bringing the awakening woods and the returning bird life to you as you drift.

It might be noted here that "meander," the descriptive noun for the geographic feature, also becomes a verb—an irreplaceable, invaluable verb when you're speaking of rivers.

*The presence of loons is also a tribute to the clarity of the water. Loons feed—underwater, on fish—by sight, and must have clear water. They have thus become a symbol of the need to protect water sources from clear-cutting and other producers of turbidity.

• • •

Limnology (from the Greek *limne*, a pool or lake) is the study of fresh water; fluvial mechanics is a branch of limnology. Fluvial mechanics tell us that even where a river channel runs straight—which, unless it is restrained, it will seldom do for long—the current within it swings from side to side, building alternating sandbars that slope first toward one bank and then the other, forcing the current to oscillate like a high-pressure hose left unattended—for much the same physical reasons. The river within the river is like a muscle in the river's body, working its way back and forth, up and down, restlessly exploring the riverbank and bottom for a way to flow faster. The flow of the river within the river will always be sinuous.

This tendency toward sinuosity works in the vertical as well as the horizontal plane. Consider the river channel as the bottom half of a slanted tube. The angle of the slant determines a river's velocity and thus its capacity for carrying sediment. If it has more load than it can carry, it dumps out the excess, building up the channel—tilting the tube downhill—which increases the gradient, velocity, and carrying capacity; if it has more carrying capacity than it is using, its excess energy cuts the channel deeper into the earth, reducing gradient. It adjusts, in other words, until there is balance between flow and load. Before that balance is reached, its channel is being cut upstream, back toward its headwaters. (In the process, floodplain meanders are often cut off from the river and left in the landscape as the familiar oxbow lakes.) Portions of the channel not yet eroded to gradient—waterfalls and other interruptions—are known as knickpoints. Knickpoints, like other geographical edges, enchant river lovers; they're where the action is. (GEOGRAPHY, says the bumper sticker, IS WHERE IT'S AT.)

Rivers never cut their way to gradient smoothly. Irregularities of rock and soil force them to gather themselves into alternating deeps and shallows. The same liveliness that drives the current from bank to bank also carves the bottom, making the pool-riffle-pool configuration so beloved of trout fishermen and fast-water enthusiasts. Riffles are high points in the bed, pools are low points. Riffles, scoured clean of small stuff by the current, are composed of coarse sediment, pebbles, and larger rocks; pools tend to have sandy bottoms because riffles act as dams to slow the flow, letting sediment sift out into the upstream pool. Surface flow in pools converges toward the center, while in riffles it diverges toward the banks—essential but probably instinctive knowledge for the fly fisherman. For reasons that are not quite understood, riffles recur in perplexing regularity every five to seven stream widths. They stay put; the deposits that form the riffle may migrate downstream during heavy flow, but they are replaced, and the shape and height of the riffle remains essentially unchanged. The thalweg, which dives toward the outside bank at every meander, crosses back again at the next riffle.

To float the Raquette in spring, we put in at the village of Long Lake in the central Adirondacks. The lake, a dramatic southwest-to-northeast slash mark on the map, is essentially a widened portion of the river, fourteen miles long and in places nearly a mile wide. The town is nine miles from the foot of the lake, which means about three hours of big-lake paddling before the shores close in and you begin to feel you're on a river. Big lakes almost always have at least a little wind on them, which can make for an unpleasant chop—or worse—and we choose our sheltering shoreline, east or west, accordingly. The eastern shoreline, after the first mile or two, is forest preserve, spotted with campsites; the western shore is accessible for some of its length by road, which means it has cottages and camps on it, but near lake's end is also undeveloped state-owned land. The western shore gives the best view of the very handsome Seward Range beyond the foot of the lake, but Long Lake, with several stunning islands to relieve the eye, presents no ugly perspectives.

The Raquette River is always our first trip of the year. There's considerable pleasure just in being back out on the water again, so the long paddle north would suit our tastes even if it weren't in a splendid setting. If time allows there's a nice little side trip up a tributary creek that trickles down to the western shore; penetrate a hundred yards and you exchange the often blustery lake for flooded marsh: bitterns, great blue herons, mergansers, the haunting "Old-Sam-Peabody, Peabody, Peabody" of the white-throated sparrow. At the mouth of that creek we've spotted the brown shape of a beaver passing beneath our canoe as we paddled.

On our most recent trip we were too busy dealing with wind to dawdle. Lake winds usually don't come up until at least midmorning, so we made an early start, but along with a bright, clear dawn came more breeze than seemed quite fair. Chris's brother Mike was along, joining her in a tandem canoe while I paddled solo. The wind continued to build all morning—mostly from behind, fortunately, but canoes want to weathervane, and following seas exacerbate the problem. Within a couple of hours we were among incipient whitecaps, and what started out as a gentle cruise had turned into a vigorous, damp, and slightly scary athletic contest.

Mercifully, however, lake eventually turned to river again, and everything calmed down. The Raquette exits Long Lake through an aisle between alder brush, flooded in spring, the sky over it aswirl with tree swallows. Here the air grew quiet, the surface still, and lake turned into flooded swampland and scratchy wilderness, where in low water the river currents must meander extravagantly to find their way down a flat-floored valley. The riverbanks are low, however, and the spring flood overruns them. You locate the channel—if you bother to do so—largely by inference.

There's little significant current for the first mile or two because the swifter Cold River comes in from the east and backs up the Raquette, amplifying the spillover. This year, frazzled from fighting the wind, we left the Raquette at this juncture and headed up the Cold for a favorite campsite at the mouth of Calkins Creek, a mile or so upstream. The state lean-to was unoccupied—once off the lake we would see only one other soul during the three-day trip—so we stopped for the night.

At dusk I went for a paddle alone, winding my way upstream through a maze of alder marsh. Pink maple buds floated on the surface, gathering into swirling rafts in the eddies, breaking apart again as the current plucked at the edges of their mass. I was cruising over six or eight feet of melted snow, clear enough even at that depth to show the uniform ripples, regular as the teeth of a comb, on the sandy bottom. Where the banks should have been I could see their waving grasses below me, overhanging darker depths: snags, logs, rocks, the diverse underwater structures where fish like to hide. The drowned banks reminded me of noodling in the Kiamichis: it was into just such darks and hollows that we stuck our unprotected hands. No wonder I'd been spooked.

The evening was still but hardly quiet, the peepers positively raucous. On my way back downstream I startled an otter that, from the noise of the splash when it left the bank, must have been the size of a small Labrador retriever. Birdsong almost drowned out the peepers. I particularly enjoyed an extensive conversation among an undetermined number of barred owls, inquiring, in the gathering dark, into each other's kitchen staffs.

I once flew over desert sand hills and was surprised to see that from a few thousand feet they looked exactly like the familiar rippled bottom of any sandy stream, viewed from about the height of a barefoot child. Nature—or physics, same thing—is fond enough of certain forms to keep repeating them. ("Nature is full of a sublime family likeness throughout her works, and delights in startling us with resemblances in the most unexpected quarters"—Emerson.)

The ripples form the same way, by the same mechanism, whether in water or air. Suspended sand itself damps the turbulence of the water or the air, reducing its carrying capacity; the sand that sifts out builds a ridge—a dune, whether submerged or not—that sticks up into and further slows the flow, which lets more sand drop out, which increases the height of the ridge, and so on. The more power behind the stream of air or water, the bigger the structures that are formed. Those pleasant little ripples under your toes at the beach are duplicated, where conditions permit, in large: in the Amazon they've been measured, a hundred feet or more below the surface, at twenty feet high and six hundred feet long.

Riffles don't move but ripples do, migrating downstream.

(So do meanders.) Ripple slopes are gentle on the upstream side, steeper on the downstream; the current carries sand up to the crest, then avalanches it down the downstream side, gradually moving the ripple toward the sea. If conditions were perfectly stable, the crest of the ripple would be a straight line, transverse to the direction of the flow. Moving water (or air) is seldom stable, however, so most crests have a long, sinuous line.

Because the particle load of a stream is moving more slowly than the water that carries it, it has a damping effect. Clear water is faster and more turbulent, more lively, than muddy water. A stream's current generally becomes progressively slower with depth, at the bottom sometimes forming a sluggish boundary layer that hardly moves at all. (Speed of the current, say the mathematically inclined limnologists, is inversely proportional to the logarithm of the depth.) I've experienced this as a swimmer, finding it possible to swim upstream along the bottom when the current at the surface was too powerful to resist.

Flow diminishes as summer wears on, not necessarily because of drought but because streambank trees leaf out and begin absorbing groundwater that would otherwise augment the stream flow. The pond behind our house, about a third of an acre in size, is mostly spring-fed, but a couple of trickling small streams feed into it from spring thaw to early August, when they usually dry up. We thought this was from diminished rainfall during the dog days, but it is actually the draw-down of groundwater. Sure enough, flow resumes after the foliage falls, whether there has been rain or not.

(In dry conditions, as in South America, the surges in flow can be diurnal. "As all these rivers have a rapid and short course, and are formed by the melting of the snow," writes

Charles Darwin in *The Voyage of the Beagle*, "the hour of the day makes a considerable difference in their volume. In the evening the stream is muddy and full, but about day-break it becomes clearer, and much less impetuous.")

The sand and gravel that streams deposit is sorted not only by the velocity of the flow but also by a process known as "fining upward": heavier particles below, smaller above, up to finest sand and silt. Some of this fining can appear uncannily methodical. Flat, disk-shaped pebbles, for example, will become stacked among other pebbles of the same approximate shape and size. They'll even be arranged in the same position, with their long dimension across the flow and their upstream edges lower than their downstream edges, so the flow can't get under and flip them out of place; over time they become assembled into tightly imbricated structures. Pebbles that don't fit get washed away, and pebbles of the "correct" (matching) size and shape are laid down in their place. A slice down through the strata of a streambank can show successive layers of silt, fine sand, then larger and larger pebbles, laid down in such orderly fashion that they look as if they were placed there by hand.

I've built a bit of stone fence on our place, attempting to sort and stack fieldstone into coherent structures. Those feeble efforts have given me a peculiar insight into the amount of sorting and reshuffling and repositioning necessary to lay down so tight a configuration. But then no wall builder will ever have the patience and energy that a river does.

Nightfall on the Raquette brought sprinkling rain and, around midnight, an electrical storm sufficiently heavy to drown out, at least for a time, the lovelorn honking of a bittern on the far shore. Dawn was clear, with Calkins Creek making small music at our feet. We were four hours from home and a thousand feet higher, putting us two weeks back into winter—which, since it forestalled bugs, was exactly what we'd hoped for. We packed wet tents and gear and set off down the Cold River again, accompanied now not only by rafts of maple buds but drifting blobs of *surfactant*, the white foam that looks like sewage but is a natural vegetable product. (It's actually a form of detergent.) Drifting along so swiftly on the spring current, the blobs kept snatching at the attention, as movement always does in the woods. Occasionally they'd pile up where upstream breezes could dislodge fluffy feathers of their foam, so that white fluff was busily drifting in one direction on the surface, the opposite direction in the air.

As we emerged from the Cold River back into the Raquette, the current began to pick up, making active paddling, other than for steering corrections, a kind of redundancy. As the flooding river spills over its banks, it cuts off necks and meanders and peninsulas, trickling off sideways from the obvious direction of the flow, enabling you to canoe the not-river as well as the river, cruising quietly over what in unflooded times would be marsh or dryland brush. Submerged reeds and grasses, bending, point the way downstream. The gurgle of the stream teases you always with new routes, secret passageways. Simply by following the current you can cut off large chunks of the river's course—but why, unless you are late in the season and the blackflies have hatched, would you want to do that?

In most graphic renditions of a loon on water, the artist draws a silver line trailing behind the bird to represent the reflection off its wake. In early morning sun on the Raquette, the flooded gray backlit woods were a collection of those silver lines, an infinity of parallel strands drawn across the water at every upwelling of current, interruptive tree trunk, or cat's paw of wind across the surface. Looking into the sun, as we followed the twists of the river, all was silvery glitter, no color at all but a light show nonetheless; then we would round a bend, the sun would be behind us, and what had been simple black and white—or silver and gray—was suddenly brown grass, green moss, yellow sand, deep green conifers. Next bend, back to black and white.

Tall, old-growth white pines, their tiers of horizontal branches pointing out the direction of the prevailing wind, tilted toward the river, their roots gradually being undermined. Similarly undercut maples leaned out over the water, lower branches making a kingfisher heaven. An occasional birch was gnawed three-quarters through by beavers, then abandoned when the top snagged. The quintessential Adirondack scene—a favorite of Winslow Homer—is a dark pool on a bend of a river, with a hillside conifer grove behind. The pool in front is shaded, deep and clear, spotted with lily pads; perhaps there's a sandbank with a crumbling turf line above it, overhanging grass and green moss, a granite boulder or two. Such a spot seems to come up every couple of hundred yards on the Raquette.

By midmorning, hills closed in on the river's course, and we began to hear the roar of Raquette Falls. The current doesn't accelerate dangerously until the last hundred yards or so, but for the hearing-impaired there's also a large warning sign on a midriver rock, reinforced nearby with a garland of twisted aluminum that used to be a canoe. Some guidebooks list Raquette Falls as runnable for experienced canoeists, but in springtime high water they are not so, by any stretch of the imagination. The falls proper are two steep fifteen-foot interruptions in what is otherwise a mile and a third of fairly sharp rapids.

(At one time there was a three-foot dam at the head of the rapids, raising the upstream water level so the *Buttercup*, an eighteen-foot steam launch, could bring tourists up from Long Lake Village to see the falls. One night in 1885, however, the *Buttercup* mysteriously sank, and the dam was dynamited. The guide business, which had been in a slump, immediately recovered.)

Most of the mile-and-a-third portage around the falls—called a "carry" in the Adirondacks—is an old tote road, broad but steep in parts, climbing a hundred feet or so and necessarily descending about twice that amount. Wood violets, trout lilies, and spring beauties line the trail, and we've often seen deer while we were hiking. The portage path leaves the river for its entire length, but if you want a good look at the rapids there's a rough streamside trail. We returned that way for our second load, shuddering a bit at the thundering force of the falls. Scared the pancakes out of me.

By sheer chance we'd begun the carry just as an annular eclipse of the sun was gathering itself, and we finished our second trip with the black dot of the moon centered in the sun's disk, flares radiating out from its circumference. Then dark clouds and rain swept in, and we couldn't tell whether the darkness was from the eclipse or the downpour. The ranger station at the foot of the falls was unoccupied, and we dodged onto its front porch for a dry and leisurely lunch; the squall passed, the sun returned, the dark spot of the moon now edging toward its other perimeter. As we were reloading our canoes to continue downstream, a New York Department of Environmental Conservation helicopter came whop-whopping up the river at about three hundred feet, opened its cloaca, and spewed a stream of trout fingerlings into the river below. Toward the river, anyway.

Where the slope of a river bottom flattens out suddenly, rebounding water throws up a standing wave, which stays in place while its water curls in the upstream direction, like a comber breaking on a beach. Fast-water enthusiasts are fond of standing waves, some of which are tall enough and predictable enough to be surfed: a boat pointed upstream can sit on them, sliding down the hill of the wave at exactly the same pace as the water passing underneath, held in place by the balance of forces (and some skillful paddling).

An underwater obstruction at the foot of a steep stretch of stream bottom can cause an *antidune* to form. The back curl of the standing wave it causes—which wave, incidentally, will be one and a half to two times as high as the obstruction that causes it—reverses the usual fluvial mechanics, causing erosion on the downstream face and deposition on the upstream. Because of this the dune that forms will migrate upstream instead of down, which is why it's called an antidune.

The whitewater folks have names for all of these phenomena: pillows, tongues, chutes, souse holes, cushions, hydraulics, haystacks. The eddy—downstream of every boulder or tree trunk or other intrusion into the course of the water, where the current curls back and momentarily reverses itself—is the most valuable tool in whitewater boating: a place to rest. When you're traveling upstream you can paddle from eddy to eddy and avoid much of the force of the current running against you. In heavy current, eddies are everything.

Mathematical relationships don't stop with riffles and pools and the height of standing waves. Meanders—the product of what one limnologist calls the river's desire to do as little work as possible—tend to have a wave length of seven to ten times the width of the river and an amplitude of eleven to sixteen times, approximately double the distance between riffles. The rare straight channel will not remain so for more than a distance equivalent to about ten channel widths. I have tried to figure out the significance of these ratios for recreational purposes, but haven't a clue. Neither have the limnologists, whose interests are considerably less narrow than mine. Nobody quite knows why they happen the way they do. They are clearly harmonic: if we could hear the river's true music, it might be playing something like Bach.

I also try to keep my exceedingly limited grasp of limnology in mind as I spend more time on rivers, but the principles seldom demonstrate themselves in ways I can recognize. It is fitting for science to attempt to understand them, but part of what makes rivers the lively things they are is their resistance to analysis. They resist management even more successfully. Occasionally we get our hands on a river that is "tame" enough—an interesting locution—to be managed. When we do, we kill it.

"The new creek bed is ditched straight as a ruler; it has been 'uncurled' by the county engineer to hurry the run-off. On the hill in the background are contoured strip-crops; they have been 'curled' by the erosion engineer to retard the runoff. The water must be confused by so much advice"—Aldo Leopold, *A Sand County Almanac.*

To be on the Raquette in springtime is to be drenched in bird-song, afloat in birdsong. (Birdwing, too, since one of the more entertaining sound effects is the looney-tunes whistle from the wingtips of darting snipes, as they whirl overhead in courtship aerobatics.) The song of the whitethroat, grace note of the north woods, echoes everywhere. Some years the backwater marshes are alive with Canada geese, whose massed takeoffs are always accompanied by a pandemonium of honking. Ordinarily we hear loons from out on Long Lake, but they were scarce this year—and would've been drowned out anyway by our bittern, sounding through the night like a two-hundred-pound bullfrog. Coyotes can be depended on to yodel overnight. You may see the vee-shaped wake of a muskrat paddling by, or a beaver. Heavy use of the Raquette drives away the otters and mink that are fairly common on wilder Adirondack rivers, but red squirrels, constantly cursing, guard nearly every campsite. Your canoe drifts over drowned trees, their submerged limbs wagging back and forth at or just under the surface. Progress is silent enough that you often sneak up on browsing deer on the shoreline—just the way my father used to promise.

A couple of miles before our planned takeout we made another side trip, a kind of guerrilla expedition, to Follensby Pond. Follensby Park, one of the largest remaining private landholdings in the Adirondacks, was scheduled to be acquired by the state shortly after our spring trip, but when we were there was still posted, with rumors of a caretaker guarding its boundaries. (Canoeists elsewhere in the Adirondacks, allegedly trespassing, have been arrested. Access to navigable canoe waters is supposed to be guaranteed by very old law, but rich landowners tend to honor this custom only in the breach. Access is a big issue in the park, resolution of which lawyers have of course slowed to an extremely muddy pace.)

Follensby Pond is the site where in 1858 a group of heavy thinkers from Boston, including Ralph Waldo Emerson, Louis Agassiz, and James Russell Lowell, pitched their famous "Philosophers' Camp." The pond is furthermore reputed to be one of the region's most beautiful, so isolated that it was chosen as the site for reintroduction of bald eagles. We wanted to see it, and if possible them. We made it well up the access creek—the posted signs were underwater, and if challenged we planned to plead stupidity—before a low earthen dam made us feel unwanted enough to turn back. Did hear a loon on the pond, couldn't see it; a pair of buffleheads, the male spectacular in its black and white plumage, teased us at the limit of binocular range all the way back to the Raquette.

Below Raquette Falls you're quickly back in floodplain meanders, and the remainder of the trip is a gentle float: sixteen miles to the highway takeout, or twenty three to Tupper Lake. The last seven miles is more or less channelized, and lined with cottages and other development (but usually has more bird life than the deeper woods: birds like action). Because powerboats have access to the lower river, there's an occasional stink and drone to put up with, but traffic is not heavy in spring. We saw only two motorboats on Long Lake and one on the river. The twelve miles below the falls are as historically rich as anyplace in the Adirondacks, but spring floods have a way of overpowering history, immersing you totally in the huge physical change that once more is transpiring in the land.

"For the first time it occurred to me this afternoon what a piece of wonder a river is," wrote Thoreau in his journal for September 5, 1838—"a huge volume of matter ceaselessly rolling through the fields and meadows of this substantial earth, making haste from the high places, by stable dwellings of men and Egyptian Pyramids, to its restless reservoir. One could think that, by a very natural impulse, the dwellers upon the headwaters of the Mississippi and Amazon would follow in the trail of their waters to see the end of the matter."

3

loving water

> There is ever a lurking suspicion that the beginning of things is in some way associated with water, and one may notice that in his private walks he is led by a curious attraction to fetch all the springs and ponds in his route, as if by them was the place for wonders and miracles to happen.
>
> —John Burroughs, "Nate's Pond"

My father liked to fish nice waters; my mother needed to get *in* them. Gwen's attraction to swimmable waters of any kind made Ralph's affinity for the Mountain Fork look like a passing whim. I never heard her express the thought in so many words—she may not have been entirely conscious of it—but the case might be made that water, and immersion in it, was the organizing principle of her life.

She was a beautiful swimmer. When she was in her seventies we both swam regularly at a local Y. Sometimes we went separately, and once I realized that she was also in the pool when I recognized not her face or her shape but her swimming stroke—from underwater, four lanes away. In my childhood there was a trophy on our mantel that she'd won as a teenager at the Tulsa YWCA, for what must have been a kind of exhibition swimming. I was unfamiliar with that kind of competition, and in later years asked her about it. All she could remember was that it was connected with lifesaving training. That part I knew about: in family legend she'd disgraced herself in one of the drowning drills when, unable to free herself, she'd bitten the instructor's thumb to make him let her go.

Think of those old black-and-white newsreels, of what we used to call bathing beauties, tucking curls up under the edge of those curious rubber helmets, complete with chin strap, that women used to wear when they swam. That was Gwen's era; that's what, in effect, she was: my mother the bathing beauty. Gertrude Ederle was her heroine. (Johnny Weismuller was mine.) Swimming must have been her first success, her first freedom when she was growing up, and I think it left her slightly water obsessed for the rest of her life. She didn't talk about it, but she so arranged matters that it was usually available. She never missed a chance to offer its freedom to Jud and me. She took us swimming whenever and wherever she could, establishing early that that was the ultimate treat, the best reward—and, perhaps, the most reliable escape.

She was an escapist, I suppose, but also an escape artist, starting from a shack on a back corner of a ranch outside Haskell, Oklahoma, in what was still, for another ten months after her birth, Indian Territory. I've seen the place: three small rooms and a summertime cooking porch, one step up from a dirt floor shanty, for a family of seven. When her father died, her mother fled to Tulsa to run a boarding house and try to get the three daughters still at home through high school. Tulsa had a YWCA, and Gwen, the youngest in the family, began to swim.

I had no real idea what the young Gwen had looked like until five years after she died, when Jud discovered a trove of old photographs. When she married Ralph Jerome in 1926 she was a slim brunette, a surprising amount of whose facial structure lives on in her first grandchild, my eldest daughter, Kate. I think she was quite pretty, but then when I was six I was certain she was the most beautiful woman who'd ever lived, except perhaps for Claudette Colbert, to whom in my eyes she bore a slight resemblance.

It was an abusive marriage, and she escaped again, moving us from Oklahoma City to Houston. Jud was ten and I was four. She must have been a fairly blithe spirit. Even after the divorce—after I was old enough to remember—she could still be inveigled into demonstrating her cartwheel on anyone's lawn, and sometimes, in party circumstances, in their living rooms. She loved to dance, although I have no memories of her dancing and never danced with her, as any dutiful son might well have done. When she talked about high school beaus she mentioned first their looks, second their skills as dancers. Ralph had both. She was stunningly prudish with Jud and me, at least in expressed attitudes, but there was a carefully hidden sensuality that we always knew was there. She did things for how they felt, I think; certainly there wasn't a lot of analysis going on. Her affinity for cool water was the strongest evidence we had. I believe she touted it to us, perhaps subliminally, in just that way: how good it felt.

It is presumptuous to attempt to imagine anyone else's sensual life, and preposterous when the subject is one's own mother, but the pleasure that Gwen took in water makes the subject irresistible. The messages were decidedly mixed. In her role as mother she hated bad language and dirty jokes, seemed to fear sexual license in any form. She apparently regarded my pubescence as a particularly unfortunate turn of events. Yet she'd share a bathtub with Jud or me until we were nearly of school age, and long after that would occasionally lapse into a kind of absentminded nudity, at least until my overheated interest in female bodies became too obvious. When I was twelve she was monumentally pregnant with my younger half-brother Dan, and got into a nest of chiggers while picking wild grapes, which left her body covered with bites. She began taking oatmeal baths to alleviate the itching, and one day, as she stood naked in the tub drying off, called me in to witness the amazing sight: her swollen breasts and her enormous tick-tight belly so spotted with chigger bites that there wasn't an unbitten area larger than a half-dollar between her armpits and her knees. It was quite a spectacle, all right.

I don't know what to make of this casualness. It seemed perfectly natural at the time—Ralph had also not been shy about nakedness, and I can't remember any hint of discomfort or embarrassment about it. Jud thought Gwen's behavior was seductive, but he saw seduction everywhere (and routinely volunteered as its victim). Between marriages Gwen had a two-year career as a divorcée, and was not without a social life, but I have little memory of it. A certain amount of enthusiastic, post-Prohibition drinking went on. A family joke had it that Gwen was the one with a ridge across her nose, from the Mason jar the white lightning came in. This was the playful view of drinking—drinking to be funny, to make things happen. Ralph's drinking, on the other hand, the cause of their divorce, was the real thing.

Gwen's true family, other than her children, were her three older sisters, Audrey, Orpha, and Peggy—ruddy, round-faced Scots with raucous, whiskey-voiced laughs. (The Jeromes, by contrast, were pinch-faced worriers.) Orpha had a daughter named Maxine who was about Gwen's age and a knockout of a brunette. Maxine and Gwen sometimes double-dated, and that was how Gwen met the stolid German oil-refinery worker, Ott Luer, who would become her second husband, to my enduring Oedipal displeasure. There were tensions in her second marriage too, but they had to do with coldness and indifference, not alcohol or sexual license. Ott and I feuded, I left home and saw little of either of them for several years. Then suddenly Gwen was fifty and newly slimmed down from yoga classes, her hair in a long single braid down her back, a handsome woman who had suddenly, strikingly, gone gray. I was a little startled at how attractive she still was. Considerably later she told me that Ott had made her quit the yoga classes: they made her horny, and he didn't like it.

She let us *go*, that was the astonishing thing. I was forbidden bubble gum (too sensual) and comic books (too trashy), which I considered brutal deprivations at the time, but if Jud or I came up with an adventure, a plan, an attempt to accomplish something or to get somewhere—usually having to do with a body of water—she'd help us as much as she could and then quietly let us go. By early teenage years we were hitchhiking all over the state, making ill-financed and unplanned expeditions to anywhere we fancied, coming and going on our own schedules. We were never even seriously questioned. Sometimes she went with us. During World War II the three of us rode dreadful old balloon-tired, sixty-pound bicycles from Houston to Galveston and back, fifty miles each way, during her week of summer vacation. We stayed in borrowed cottages, living on hot dogs and soda pop; the reward was swimming for a couple of days in lukewarm, milk chocolate–colored water on an inlet off Galveston Bay, very near the spot where, decades later, LBJ would build his space center. Diving into that opacity was an act of faith; alligator gars, sinister-looking if you could see them but in fact harmless, would brush our legs as we treaded water. Ott wasn't along, and we had a terrific time. I suspect that's one reason she was so liberal with our freedom. She knew how badly we needed to get out of the house.

A few years after Ott died, when Gwen was in her early seventies, she rather abruptly moved in with Chris and me, an arrangement insufficiently negotiated beforehand. It was not a congenial solution, difficult for both women, who had different styles and didn't know each other very well. After four years Gwen announced with the same abruptness that she was moving out again. Chris and I were ashamed of our relief but relieved nonetheless. Her best friend's husband was dying, she needed help, so Gwen moved back to Texas and took an apartment in the town of New Braunfels—where we'd lived for several years, where I'd gone to high school—and lived there for the remainder of her life.

A few years after she died I had the chance to check out one of her favorite swimming stories. Shortly after World War I, in her adolescence, she'd made a summer trip to Florida, and had gone swimming in that state's crystalline freshwater springs. For an Oklahoma farmgirl, for whom summer had always meant inescapable, un-air-conditioned heat, it was an ecstatic experience, one she would talk about the rest of her life. By the time I was six, and swimming myself, she had convinced me that I must eventually go try those springs, but I didn't get around to it for another fifty-four years.

Springs are magical places, where the earth for no apparent reason decides to give us fresh water. The first one I remember was in the Kiamichis, across the river from Coleman's place, tucked away in a shady hollow, bubbling cold in hottest summer. It was tiny—three or four feet across, nowhere more than a few inches deep—and collected mosquitoes, ticks, and snakes. It wasn't exactly a place to linger, but it was enchanting to visit. One drank from it by kneeling, kissing the surface as wood sprites are supposed to do. In my twenties I lived for a while in western Texas, in an area that gets about eight inches of annual rainfall. It is a region originally made habitable only by grace of giant springs—at Sweetwater, Big Spring, the names tell all—that lined the desert below the dreaded Llano Estacado. By the time I saw that country, most of the springs had dried up or were pumped off into town water supplies, but the relics were still there, ghostly great dry caves that conjured wagon trains, Indian ambushes, desperate situations in the surrounding wasteland. As Chris points out, the earth around a water source is always a palimpsest.

Florida is perhaps overblessed with them. Its soil is sand from ancient seas, sitting on a limestone pan made up of ancient reefs. Acidity in the rainfall, some of it even from natural sources, eats holes in the limestone. Water percolates downward, undermines the sandy soil, sucks down large chunks of it to form the state's thousands of sinkhole lakes, and, here and there, bubbles back up again, purified, in the artesian springs that Gwen had so enjoyed. There are hundreds, I suppose; I circled thirty of them on the highway map before I stopped counting, and visited two of the more famous—Homosassa Springs, where there are manatees, and Silver Springs, where the old Johnny Weismuller Tarzan movies were filmed—before I succeeded in immersing myself. Both had long ago turned into tourist traps.

Homosassa is an impressive hole in a river, a huge, dark blue cavern with substantial numbers of quite large fish stacked like logs in the current, disappearing eerily into the darkness below. From it a good-size river simply *starts*, where there is no discernible reason for a river to be. Gentle, cowlike manatees drift at the surface, grazing on heads of lettuce provided by their caretakers. Silver Springs is similar, although considerably larger (and cleaner), and was curiously recognizable, perhaps because as a kid I'd seen every one of the Tarzan movies, had virtually memorized them. Seventy years ago it had been open for public swimming, and might well have been one of Gwen's magic places. Now it was inhabited by glass-bottomed boats and surrounded by amusement park kitsch. I petted a giraffe on the way out.

A friend at the Nature Conservancy had suggested Juniper Springs, swimmable and canoeable, in Ocala National Forest, which turned out to be what I'd had in mind all along—a small, nearly deserted state park in the middle of national forest, with minimal development. The natural oval of the spring has been walled with stone and surrounded by a concrete deck, giving much the feeling of a rather antiquated municipal pool, albeit with a white sand bottom. Outflow was over a small dam, beside a partially restored mill house, but below the dam the stream had been left more or less wild. It was February—I'd failed to duplicate the hot summer conditions that had so colored Gwen's experience—and the place was quiet, calm, almost deserted. Three teenagers weren't so much swimming as plunging in and climbing quickly back out again to loll in the weak sun.

I got my gear—I'd brought along mask, fins, and even a shortie wet suit in case the legendary spring water was uncomfortably cold—and changed, shivering a little at bare feet on the bathhouse floor, wondering if I were, well, *vital* enough for this experience. By the time I'd dressed I had the pool to myself, and slipped on in. Not bad. Just fine, in fact, the wet suit proving an unnecessary timidity. Delightful. What had all my dithering been about?

I snorkeled along on the surface long enough to stabilize my breath, then swam down toward the bottom, looking for the source. Maximum depth was about fifteen feet, in a small cave leading off to one side, drifted two-thirds full of sand. I couldn't see very far into its darkness, but it didn't seem to go anywhere. The inrushing water bubbled through the surface of the sand, nippling it like pea soup coming to a boil. I could plunge my hand into the bubbling sand without disturbing it, and feel the swirling pressure of the emerging water: not a fire hose but a powerful, steady flow.

I came up for air, dived again, cruised upside down a foot or two off the white sand bottom, dolphin-kicking lazily, looking up at the silvery underside of the surface above, the top of my head slithering through waving blue-green eel grass. No fish to speak of, but I had a hunch they were around. If I were a fish, I'd want to be here. I swam the perimeter, enjoying the sights, trying to understand what was so familiar about the startling clarity, the color scheme, the waving plant life. Then the memory clicked in: these springs reminded me of the Comal River in New Braunfels, a spring-fed clearwater river that surges up from the base of the Balcones Escarpment. We'd lived near the Comal in my adolescence—Gwen had moved us there from Houston—and I'd swum every available inch of it. Same colors, same languid currents, same abundance of plant life. Same superb swimming.

Well. I couldn't quite say the experience—the February version, unfortunately, of what Gwen had treasured so for all those years—would top the list of things to tell my own progeny they must someday do. I'd swum in enough muddy rivers, however, and survived a sufficiency of southern summers, to understand that contrast is all, and alongside what Gwen had known, these places must have been simply stupefying. I saw that clearly enough. I think in the process I even understood one little piece more about her. Ecstasy may be too strong a word for what she sought, but something like it was implied by her restless lifelong search for ways to repeat that vividly remembered pleasure.

The clarity of the water reminded me of an ecstatic swim of my own, in 1950, in the headwaters of the Comal, at about one A.M. on a sweltering hot full-moon Texas night, with a girl who was soon to become a lost love. Jud was also along, with a new young wife, and another couple, the male half of whom I'd pulled unconscious from the same river a few days before. He'd hit his head diving and was alive to go swimming with us again only because of my girlfriend's talent for what was known in those days as artificial respiration.

The near tragedy sobered us for a day or two, but we were very young, it was deep summer, and the Comal continued to beckon. On the night in question, after a certain amount of beer, we trespassed into a privately owned stretch of the river and slipped in quietly, squelching shrieks at the frigid water, shushing each other in giggling conspiracy. The escapade had been my idea but had gotten slightly out of hand; I was apprehensive, still frightened by our friend's narrow escape, not at all sure anymore that risk and swimming were a good combination. Until I slipped under, pushed off, glided through the silver water, watching with blurry eyes the dark shadow of my own naked body drift across the moonlit sand below me, and the confusing world reduced itself to black and white, to cold and hot, and some unnameable voice whispered in my head, as I slid along under the surface, *alive! Alive!*

4

the application of force

> When asked if he had ever tried out the stroke in the water, Sir Nicholas replies: "No Sir, but I swim most exquisitely on land. I content myself with the Speculative part of swimming, I care not for the Practick. I seldom bring anything to use, 'tis not my way."
>
> —Charles Sprawson, quoting from Thomas Shadwell's *The Virtuoso*, in *Haunts of the Black Masseur*

Until a few years ago my experience with canoes was limited, but the more time I spend in them, the more I've come to enjoy the act of paddling. Not obsessively, I think: I don't love the fifth consecutive hour on a windy lake. I'm as ready as anyone to quit early and loll around the beach for the rest of the day. But I will go for a paddle for the pleasure of it. To make a canoe go smoothly through the water is a curiously satisfying thing to do, a form of physical effort that my aging frame finds congenial. I find it as pleasurable as some find dancing, or other ways of using the large muscles in rhythmic motion. "Primitive" societies have used similar techniques to achieve various levels of contentment, up to and including trance states, throughout recorded time. That paddling a canoe so often takes you through beautiful country at the same time it is giving you pleasurable exercise is pure bonus.

This affinity for paddling is obviously a direct outgrowth of the way I feel about swimming. Since early adolescence I've been guilty of staring at good swimmers. It gives me a surge of pleasure to see someone swimming easily and well. I feel it in my shoulders and hands, the same places one feels pleasure in watching a hawk work the air. I feel about the swimming stroke the way golfers must feel about the golf swing, or baseball players about the batting stroke. It is a skilled application of force.

This beguilement has now turned to the paddle stroke. Most canoeists are duffers like me, but occasionally I see someone paddling in a fashion that is clearly more efficient than the way most of us go at it. When I do I am enthralled, trying to see into the stroke itself, to see how this person makes this happen with this strange appliance.

Similar physics apply. Both swimming and canoeing require that you catch hold of a bit of fluid, in effect, and pull on it to achieve movement. This medium is heavy, resistant, sluggishly obstructionist, yet slippery and hard to catch. Both strokes therefore put a high premium on efficiency. What a good stroke tells you is that the swimmer or paddler has put in a lot of miles, quite often as a racer, for whom the fatigue of long training has honed away inefficiency. Dragging yourself over distances with your arms is tiring, whether you are in the water or on it. You want to take as few strokes as possible, and to get maximum return from each. You achieve this return by finding the most accurate way to apply power. The mechanics are seldom as obvious as they seem; strokes can be very subtle. The most effective ones are the most graceful, the most satisfying to watch as well as to perform. Grace is always the product of good physics.

I once tried, for about fifteen minutes, to learn to ride a unicycle, a malevolent little machine that does not want you on it, never mind any fancy tricks like actually applying force to its pedals. The more of my weight I tried to entrust to it, the more quickly it shrugged me off. It positively squirted out from under, throwing me to the ground in new ways each time, hacking at exposed body parts as I went down. It was as if it specialized in inflicting pain—quickly, remorselessly, ingeniously applied. Fifteen minutes was enough; I never came anywhere near riding the damned thing.

Several decades later I was given an opportunity to try board sailing, or, as it's sometimes called, windsurfing. Board sailing, like unicycle riding, has a reputation for a long, flat learning curve. The board you sail on, I quickly found, might indeed be regarded as a slightly more forgiving unicycle. What it throws you down on is softer than the ground, and the hardware it flails you with on the way down is not as lethal as that of the unicycle, but flail it does, and fall you do, after each fall to climb back up on the board and haul the sail upright again. The sail might be considered the handful of eggs you're supposed to juggle while you're riding the unicycle.

My introduction to the sport took place on Aruba, an otherwise devastated Caribbean island that features, along with a standard assortment of tacky high-rise hotels and garish gambling dens, near-perfect board sailing conditions: miles of nonfrigid, waist-deep water and absurdly consistent offshore winds. My adult son, Marty, joined me. He had no more board sailing experience than I did, and I was fifteen minutes ahead of him on unicycles.

It took us some time to acquire sufficient balance and stability just to stand on the board, which seemed unnecessarily tippy. Beginners start on a large board for maximum stability, with a small sail for minimum harassment from the wind. One is encouraged to switch as quickly as possible to a smaller board and larger sail, however, because for thrills the big-board, small-sail combination is about as exciting as riding a tricycle, although considerably more difficult. Unfortunately, the smaller the board, the less stability, and the bigger the sail, the harder you are yanked around by the wind. Being yanked around by the wind is exactly what the experts are after, of course.

At first attempt, then, board sailing is a little like balancing on a bongo board while trying to repack a parachute in a windstorm. You're standing on what feels like a kind of greased ball, and holding onto a hugely unwieldy lever that is inserted into the wind. When the wind presses on the lever, it either yanks you off the board or makes the greased ball you're standing on go scooting out from under you.

The sail is anchored to the board by a U-joint that allows it to flop with spineless irresolution in any direction. Instructions for its operation are simple. Lean the mast toward the rear of the board to turn into the wind. Lean it toward the front of the board to steer away from the wind. Lean it away from your body, or closer to it, for balance. Increase or decrease the angle of the sail to the wind to control your speed.

Simple to say, not simple to do: actually you are using two hands to do three things at once with the sail, in at least six different directions, although in practice the number of directions is infinite. Then there are the two feet, with which you not only attempt to maintain balance but also help steer: four appendages working simultaneously, differently, appropriately, pursuing control. Master the sail while you are mastering the board. Helicopter pilots come to mind, muttering about rubbing the belly while patting the head.

Paddling a canoe certainly looks simple enough: you stick this thing into the water, pull on it, and the canoe moves forward. Hammer and nail. (I thought carpentry was simple too until I actually tried building things.) But a stroke that drives the canoe forward also steers it away from the side on which you're paddling. To make the canoe go in a straight line you have to keep switching sides, which is inefficient. To become efficient you have to find a way to compensate.

It is perhaps not so simple. You're sitting beside an available half of a hemisphere—a hemidemisphere, or one-quarter of a universe—in which to work the paddle. Where you put the paddle into the water, and where you take it out, and what path you pull it through, and the angle of its blade in relation to the centerline of the boat, and several other possible variations I haven't thought of yet, all have a direct result on the efficiency of the stroke. There is available, on every stroke, something like a quarter of an infinity of possibilities for the application of force.

Yet good paddlers almost all paddle in approximately the same way. What's more, particularly in swift water or other demanding conditions, good paddlers often don't make strokes at all, in the sense of sticking the paddle into the water and pulling it out again. Canoe and paddler are both more stable while the paddle is in the water; in difficult water good paddlers often just leave it there, applying force to it when and in the direction needed.

The swimming stroke works on the same principle. Swimming is, in effect, sculling. You should work the water with your hands as easily and naturally as you would wave smoke away from your eyes. You reach and "pull," but the pull is with the arm; the hand is not so much pulling as it is sculling about in search of purchase. Inexperienced swimmers try to pull in a straight line beneath the body, which is not only almost impossible to do but would be ineffective if you managed to do it. In a good swimming stroke the hand moves in an S-shaped curve, sliding downward and outward at entry, then sweeping back under the body, finishing with another swirl to the outside again near the hip. Out, in, out again.

This cursive stroke used to be explained as a search for still water, on the theory that once you got a handful of water moving, it no longer served as a useful anchor against which to apply more force. Then researchers discovered that a skilled swimmer's hand actually comes out of the water ahead of where it went in: the hand functions as a propeller blade, generating forward motion as it slices through the water in exactly the same way an airplane wing generates lift as it slices through the air. That sculling S-curve is not so much grabbing hold of water and pulling against it as it is cutting through the water to generate lift, making the water itself pull the swimmer forward.

In search of efficiency, propeller blades are precisely pitched in angle of attack. Since we can't rotate our hands like propeller blades (God, so far as we know, doesn't do U-joints), the S-curve is the compromise path that ideally allows the best available pitch of the hand to be combined with the best leverage for the muscles and joints of the arm, hand, and upper body. It is not easy to find the most effective pitch. In fact the power is usually a few degrees off the line along which you seem to be pulling. You have to learn to feel out and compensate for this deceptive vector. You try to sense the tiniest deflections from dead ahead, the perfect straight line for forward motion. The closer you can come to finding the pitch that puts the power on that straight line, the faster you go with the same expenditure of energy.

When you get it right, it's as if the quality of the water changes. It gets solider; you find the same firm resistance, the same possibility, throughout the path of the stroke—because now you know just where to reach to find and apply power.

Water slows; hurrying, in it or on it, requires an enormous expenditure of energy. Canoe racing, in this sense, is an oxymoron, although of course people insist on doing it. If speed is not necessarily the point, however, efficiency always is. I love free diving—swimming deep with mask and fins—for the sheer physical pleasure of it, but also for the chance to watch fish. The "swimming stroke" of a fish, if such a bizarre notion is possible in a creature that lives by moving through the water, is an object lesson in efficiency, a rueful reminder of how out of place we are in their realm. To chase a fish, even with fins, is to be a tree chasing a hummingbird.

Beginning board sailors reverse direction by falling off, climbing back up, and starting over again headed the other way; experts turn around by an exquisite but complex dance that scarcely slows the board. Learning to turn around is the first step out of beginnerhood. There are two ways to do it. You tack by turning the board directly into the wind, walking around the mast to the other side of the sail, and sailing off again in the new direction; or you jibe, by turning the board downwind, letting go of the sail so it can spin around the mast in front of you, stepping to the other side of the board, and sailing off again in the new direction. Easy as kiss my hand. Some find the tack easier, some the jibe. Tacking—working upwind—has the advantage of keeping the board from drifting farther out to sea; jibing has the advantage of not requiring so much strolling around on the board, which is usually when you fall. Both help prevent you from disappearing over the horizon, necessitating an embarrassing rescue.

The learning curve was indeed flat, but Marty and I began to make progress. Even in the first day's trembling successes, the little moments when things would go modestly well for sometimes fifty or sixty yards at a time, I could see what it might be like when you were comfortable on the board, when you were stable enough to take the pressure on the sail and use it to make the board respond, doing what you wanted it to do.

The trick to learning the unicycle, which no one told me at the time, is to use crutches. The only available crutch for board sailing is the wind. It is your stabilizer, your outrigger, your friend, and you must learn to lean on it. As friends go, unfortunately, it is not all that reliable. In fact it is a flirtatious and whimsical bitch goddess, stalling suddenly (dropping you on your back), puffing suddenly (face plant), usually feinting in one direction just before it knocks you over in the other. Imagine how difficult skiing would be—or walking, for that matter—if gravity came in puffs.

Mariners have been living with this whimsicality, on a much more serious scale, throughout human history, and it should come as no surprise to anyone who has ever stepped outdoors. But in a lifetime of open-air activities, wind had previously served only to make things slightly easier when it was behind me and considerably harder when it was not. Wind doesn't make board sailing easier, it makes it possible, being engine as well as crutch. Learning to live with it took longer and was more frustrating than any other such activity I've tried.

I sit with my coffee at dawn on an exposed point of a small lake, watching the mists hang over the surface of the water. Over my shoulder I hear an indefinable rushing noise, and then a merganser comes streaking by, flying so low that its downstrokes almost touch water. The air is still enough that the whirling vortices spilling off the duck's wingtips cut a visible wake through the mist, leaving a line I can see long after the bird is out of sight. I am struck by the pure phenomenality of it, the way the physics happened.

It is possible to come at these watery activities with more utilitarian motives than mine. Most people, I assume, regard swimming as either a means of transportation—and self-rescue—or a frolicky way to cool off. Canoeing is a cheap, quiet way to go fishing. (Fishing, I think, is an excuse for going to water in the same way that competition is an excuse for physical exercise.) I have an athletic bias, and my attraction to both swimming and canoeing is not strictly practical. By "athletic" I mean only a sharp interest in the physical aspect of things. I am prejudiced toward a mechanical, ramps-and-levers approach.

That bias sent me off on a long and circuitous road before it led me back to water. I wandered through skiing (the on-snow version, which is another way of toying with the physics of water) and automobile racing and competitive swimming, experimenting, more or less delightedly, with their complex ways of working the world. In skiing, for example, the horsepower comes from the angle of the slope, a.k.a. gravity, and it applies its force to you, rather than the other way around. Your job is to direct its flow—down through your body and your skis to the snow—in order to go where you want to go at the speed you wish. The obsession is with control.

In automobile racing the obsession is with getting power to the road. You can have as much power, in effect, as you want. Its technology has been thoroughly explored and few mysteries remain; increase the size of the power plant (or the speed at which it operates, or the amount of fuel that it can be made to burn) and the available power goes up to match. "Available," however, is not "applied": you still have to get the power to the road. The sophistication now devolves to suspension systems, to maintaining contact between the driving wheels and the road surface so that the power can be applied. Fancy technology comes into play: computer-controlled shock absorbers, exotic tire compounds, aerodynamic body configurations that "fly" the vehicle, and thus the tire surface, more firmly against the pavement. It is in that subtle juncture between the vehicle and the medium on which it operates that the problems get exquisitely complicated. The point at which traction threatens to break down is the point where interest lies: where the rubber meets the road.

The canoe is a remarkably efficient way to move heavy loads with low effort. Its efficiency results from low drag on the water, a combination of streamlined shape and shallow draft. These same characteristics make it excessively steerable, which is fine for dodging about in whitewater but a positive hindrance in straight-line travel. To go straight ahead without switching paddling sides requires a stroke that steers and drives the canoe forward at the same time. At summer camps they teach the "J-stroke," a straight-line pull with an inefficient steering correction tacked on at the end. There are an infinity of more efficient strokes to be had, by subtly varying the angle of the paddle blade to the straight-ahead line of travel of the canoe. They're all awaiting in that hemidemisphere of available energy right there beside the boat. All you have to do is find them.

In swimming you seek out subtle angles of the hand in search of more power; in canoeing you seek out similar angles of the paddle in search of less intrusive steering control. In swimming the water seems to get more solid; in canoeing the water seems to feed energy back into the paddle. In either case you begin to know where to reach to find power. On most long canoe trips there comes a time when that happens. The paddling stroke seems to fall into place, one's sensitivity to and control of the water becomes peculiarly heightened, and playing with the paddle against the water becomes a sensuous game instead of an arduous chore. When that happens you can't wait for each new day's paddle to begin.

Between board sailing sessions I was reading *The Mauritius Command*, by Patrick O'Brian. In it he describes two square-rigged men-of-war driving through the sea "as though joined by an iron bar." After we'd gotten stable enough to enjoy steady sailing, Marty and I would occasionally find ourselves in that mode, in tandem or side by side. Even with my infant skills I'd find myself trying to wheedle out of board and sail an edge in speed, trying to pull away. It was exhilarating. No matter how fast we seemed to be going, though, we'd occasionally hear what sounded like a motorboat overtaking us, and some more experienced sailor on a tiny board with a huge sail would come blowing by with a twenty-knot differential, leaving us with wistful looks on our faces.

The object is to apply force to the board in order to drive it through the water. You acquire that force by reaching up and grabbing hold of the energy of the wind. You grab it with the big, unwieldy net that is your sail. Before you can do that you're in a very tippy sailboat; once you learn to do it you're surfing with a motor on the back. Then you can start hanging your weight from the boom and kicking the board around, pushing it forward, driving for speed. The vicious little unicycle turns into a grand prix racer—one you steer with your feet. Grabbing the wind, you feel as if you're holding a huge ball of power in your hands. You can feel it passing directly through your arms, down your trunk and legs, out your feet and into the board. It is very satisfying.

I have nothing against speed, but I caught an intriguing glimpse of something else that board sailing could be. In front of the board-sailing shop was a stone jetty enclosing a small rectangle of calm water. On our last day Marty and I stopped to watch two local youths engage in mock combat in that tiny space. They were working air so light I couldn't believe their boards even moved, yet they cruised back and forth like skaters doing school figures. They danced around their boards from one impossible pose to the next, making passes at each other that reminded me of nothing so much as bullfighters' veronicas. They did it almost entirely by manipulating the sail, pushing and patting it into place, using it the way a tightrope walker uses a parasol. They could do it because they clearly felt the air in a way we hackers had not yet learned. They had a superior sense, always, of exactly where the wind was.

The balcony of our hotel room faced west, and Marty and I would install ourselves there in the evening, as one is impelled to do in the Caribbean, for each day's glorious sunset. Terns, gulls, pelicans, and frigate birds would be working the evening shift. I fell into the usual mistake of trying to identify them, naming instead of watching. That was a pity: an hour before, I had been out there trying to catch that air for myself, the memory of the effort still in my shoulders, my fingertips; here were experts, demonstrating how one might really work the wind, and I was paying attention to the wrong thing.

Darkness would then fall with the tropics' legendary suddenness, a phenomenon that everyone always talks about but no one had ever explained to me. Angle of the sun, we decided: in temperate climes the more sharply angled sun takes longer to traverse the horizon, effectively lingering just above and then just below it; nearer the equator it plunges straight on down—near vertical whatever the season—and positively yanks that velvety tropical darkness over your head. I got considerable pleasure from thinking about that. It made me realize where we happened to be on the planet, which I always find an intriguing thought.

When the sun was gone we would usually go for an evening swim, but on our last night I demurred, still sodden from the day's duckings. Marty did want to go, and I walked with him down to the beach under a purple bruise of a sky. On the far horizon, a few miles offshore and perfectly centered in the sun's fading glow, glimmered the silhouette of a brightly lit cruise ship, reminding me of the ocean liner scene in *Amarcord*, Federico Fellini's loving reconstruction of childhood memories. Well, I thought—enjoying the sight of my grown-up son swimming out into the last light of the evening—we non-Fellinis get to have some great memories too.

Among them will be catching energy out of the sky and feeling it flow through me and into the board and thus into the water. The pleasure from these sensual sports comes, I think, from becoming, however briefly, a medium for the application of force. Board sailing transforms you, in a way I'd never experienced before, into a link between the source of power and its point of application: between the atmosphere and the surface of the earth.

5

the physics of flight

> In youth, before I lost any of my senses, I can remember that I was all alive, and inhabited my body with inexpressible satisfaction; both its weariness and its refreshment were sweet to me. This glorious earth was the most glorious musical instrument, and I was audience to its strains.
>
> —Thoreau's Journal, July 16, 1851

I know of no satisfactory explanation for why some of us have so strong a drive to immerse ourselves in a life-threatening foreign medium for which we, unlike other vertebrates, have no instinctive aptitude. To go into water is to be an intruder. It is not our medium. Maybe that's the appeal, the thrill of becoming the Other. Yet we originated there, or so science, and perhaps the unconscious, would have it. Maybe that's why its power over us seems so strongly to do with memory. Maybe we are trying to remember when we were not the Other.

I've come close to drowning twice. The first time I was six years old, taking lessons at one of Houston's municipal pools. I remember nothing about learning to swim, remember only the locker room—fifteen or twenty other naked, shivering little boys—and the beauty of the smooth turquoise surface of the water each day before we jumped in and messed it up. I also remember when, near the end of the two-week session, they let us explore the deep end. A fat kid panicked there, grabbed me, and took me under with him.

By then I was already more comfortable under water than on top, but he caught me on an exhale, and I had to swim deeper—a tip Jud had thoughtfully passed on ahead of time—to get away. I still remember quite vividly the slickness of his skin as I struggled to get free, the rapidly building pain of the respiratory drive, the relief I felt when increasing depth itself peeled him away. I was back at the gutter in time to watch his rescue. He was badly scared but unhurt. It was definitely frightening, but for some reason I wasn't traumatized by the experience. Formal instruction taught me what it's like to swim, and what it's like to screw up when you swim. Avoid floundering fat kids.

Swimming became an obsession, for which our postdivorce move to Texas had played into my hands. As Gwen began struggling to make a new life, she found it helpful to stash Jud and me in summer with her more affluent sisters. Aunt Audrey, the eldest, was a golfer, and her country club pool became our baby-sitter. Aunt Peggy had a rich neighbor with a large backyard pool and sons our age, with whom we formed a quick, firm alliance. I was forbidden the deep end until I could swim the length of the pool, and I remember the day I did it. I crawled out, stood up, and vomited strawberry shortcake all over the deck. I had a mess to clean up, but I'd earned permanent admission to the big boys' end. Swimming was all I wanted to do. I practiced, whenever I got the chance, what I'd seen in the Tarzan movies: swim that heads-up crawl with the make-believe knife in your teeth, grab the inner-tube alligator that threatened Jane and Boy, twirl over and over in the water as you stabbed it in the belly. The twirling taught me how to keep water out of my nose.

A major part of that obsession was climatological, Houston in summer being a descent into hell: temperatures in the mid-nineties with humidity to match, air-conditioning unknown—before World War II—except in movie theaters. Swimming was the only entertainment cool enough to allow vigorous activity. Unfortunately the only local natural water was in the bayous, in those years open sewers about whose horrors we were so constantly warned that whenever I got splashed with bayou water, as would occasionally happen, I would shudder and feel faintly sick.

During the war Ott moved us even deeper into Texas's mosquito-infested coastal lowlands, to the brand-new town of Old Ocean, where he'd taken a job at an aviation gas refinery. Old Ocean, a part of the war effort, consisted of the plant, some oil storage tanks, and a hastily erected prefab housing development for the refinery workers. Only two swimming holes were available. One was a muddy stretch of the San Bernard River, which offered lukewarm brown water, shade, and a good rope swing. The other was a place we called the Canal, a large ditch—thirty feet across, in some places eight or ten feet deep—that had been bulldozed out of fresh mud behind the refinery, for God knows what purpose, probably industrial waste. We swam in it anyway. We'd emerge with a thin coat of mud that we tried to scrape off before it dried into finely crazed enamel—towels being beneath consideration, as were bathing suits, since the only ones who swam there were males in early pubescence.

If the swimming was poor I didn't miss it as much as perhaps I might have, because at the time I was in the grip of another branch of fluid dynamics. I'd begun building model airplanes, gliders in particular, and was as fascinated with the possibilities of air as with those of water. The obsession had me lying for hours in a field of alfalfa behind our house, out of sight of anyone, watching buzzards soar in the southeast Texas sky, trying to understand how flight really worked. When I wasn't buzzard watching I was hanging around a Piper Cub that one of the refinery workers parked in a field a mile away. I did so hoping to watch the owner fly, and perhaps by sheer diligence to cadge a ride. That patience paid off twice, the first time with two laps of the field with a touch-and-go landing in between, the second with an introduction to aerobatics that temporarily cured me of my fever for getting airborne myself.

So did another experiment, one that developed by chance at our junior high school in the town of Sweeny, four miles away. Sweeny's student body was a fairly tough crowd: chewing tobacco was forbidden indoors; the dress code did not require shoes. We played the organized sports with considerable enthusiasm, but mostly I remember a kind of random violence—fistfights, hazings, group assaults. School represented captivity, and since we couldn't actively resist our captors, we must have chosen to torment each other instead.

One April the power company installed a new line pole on the school playground, attached the requisite guy wire at the top, and then for some reason left without anchoring the bottom end. Forty feet of extra cable curled on the ground, challenging adolescent imaginations. The cable was too stiff for jumping rope or lassoing games. We tried swinging around the base, maypole fashion, but you couldn't get up any momentum with all that extra cable dragging behind on the ground. Then some ingenious kid fetched a ladder and leaned it against the pole, and a new sport was invented. A designated flyer would climb the ladder with a wad of rags, wrap the rags around the cable to protect his hands, and hang on for dear life. Everyone else would grab the loose end of the cable and set out running away from the pole, as if trying to launch a large kite.

Imagine a misfiring slingshot in which the stone sticks in the sling. The cable haulers would reach the end of the cable, twanging it taut; the kid on the ladder—the pebble that didn't clear the slingshot—would be yanked dramatically off the ladder and high into the air, would vibrate for a couple of sine waves, and then begin to slide down the cable. If the gang on the end of the cable kept the tension on, or even managed to give it a little extra whip, the vibrations would increase. If the flyer could be flung off at launch or in mid-cable, score one for the gang on the end. If the flyer managed to hang on, he slid exhilaratingly down and crashed into the crowd on the end, after which everyone lay around on the ground laughing and cursing and spitting tobacco juice. Score one for the pebble.

When my turn as flyer came I survived launch and early vibration, and enjoyed the flight part. Then I thought I might avoid the violence at the end by landing on my feet, but misjudged the amplitude of the flight path. As I reached ground level my feet grazed the grass on a downswing, my body swung forward on the upswing, and I came back down, with undiminished speed and force, on my coccyx. Not wanting to participate further at that time, I went over to the baseball bleachers to wait for the pain to stop. I probably would have remained there, fighting tears, until after the tardy bell for the noon recess, but there was a pickup baseball game in progress, and that was the day my friend Logic

Kuban caught a line drive in the testicles and vomited on third base, which got us all laughing so hard I forgot about the pain in my tailbone.

Old Ocean was close enough to both Galveston and Corpus Christi for weekend trips. I loved tumbling in the surf, hated the stinging jellyfish, but never really came to associate the tepid, milk chocolate–colored Gulf of Mexico with swimming. I was given only one other significant escape to water during the war years. The summer Jud turned eighteen—near the end of World War II—he came home from Oklahoma University to await his imminent draft notice. He quickly found himself even more uncomfortable sharing a house with Ott than I was, and proposed that the two of us go off on a camping trip, organized almost entirely around a swimming hole in central Texas called the Blue Hole. I was twelve. We stuffed knapsacks with canned goods, blankets, and a Boy Scout cooking kit, hit up Gwen for a ten-dollar bill, and set out hitchhiking.

What I remember about that was mostly a haze of fatigue, heat, hunger, and thirst. We'd stand, thumbs out, in searing sun for what seemed like hours; when we got a ride we'd crawl into some stranger's oven-hot car and I'd fall instantly asleep. Jud, a master of budgeting, strung out the ten dollars at a rate that kept me convinced I'd faint from hunger. We slept on blankets on the rocky ground and woke in pain. Our primitive approach was nearly enough to put me permanently off camping.

Except for the swimming. The Blue Hole was near Wimberly, a wide spot in a chilly, cypress-shaded creek that left us shivering and blue-lipped in the Texas July, which was exactly what we'd come for. Its main feature was a swing that offered very long flights over deep water, and we spent most of the daylight hours inventing new aerial didoes. There were no lifeguards. In fact most of the time we had the place entirely to ourselves, resenting any tourists who showed up—unless they included other young males and we could, literally, show them the ropes. It was an amazing adventure for a twelve-year-old: essentially we owned a swimming hole for a few days, in it from dawn to dusk; at play, set loose, beyond adult control.

The Blue Hole is a feature of a piece of physical geography that was soon to warp my history permanently in a new direction. The Balcones Escarpment is a limestone cliff that slices across Texas for a hundred miles or so on a diagonal just west of Austin and San Antonio. Below the escarpment is fertile farmland, above it LBJ's hill country: cedar brush, arid hardpan, goat ranches. At the cliff's base artesian springs bubble up, the catchment of the hills filtered through miles of prehistoric radiolaria. A wonderful big swimming pool in Austin, called Barton Springs, is wholly supplied by this hydrology. Eighteen miles southwest of Austin, the San Marcos River is clear enough to support glass-bottomed boat rides and an underwater theater, à la Florida, as a tourist attraction.

Another twenty miles along the same line is the town of New Braunfels, tucked butt up against the escarpment. There the Comal River rises, winds scenically for two short miles through the town of twenty thousand, and, still within city limits, empties into the considerably larger Guadalupe River. The Comal remains at a year-round temperature of about seventy-five degrees. That meant that in midsummer, when the air-to-water temperature differential could be thirty degrees, the Comal felt breathtakingly cold, yet in the mild south Texas winter—so long as there was no wind—it was still swimmable.

We moved to New Braunfels when I was sixteen—at Gwen's instigation, I realize now, and for the water, although that was never the spoken reason—and I found what I thought at the time would be my life's work: river rat. I lifeguarded when I was forced to work and capered elsewhere in the Comal when I was not. Lifeguarding meant the Landa Park pool, a diverted three-hundred-yard stretch of the river with diving boards; preferable to that was a place called Dittlinger dam, still the second-best swimming hole I've ever seen. An abandoned grist mill, it was in effect a stone building through which ran a whitewater river. It featured underwater passages, spillways, ledges at various heights from which to dive. It was far too dangerous for general recreational use, so my friends and I learned every inch and crack, swam it all, and taunted visitors with our skills and our knowledge. What else would you do when you're a kid at a swimming hole?

Landa Park, on the other hand, gave me my first extended exposure to diving boards, and I quickly fell in love with that sport too. Diving is recreational physics, a means of pure experimentation with mass, acceleration, angular momentum, moments of inertia. The end of the board is point A, the surface of the water is point B, and the challenge is to fill the space in between not with beauty—to try to be beautiful is automatically to become ugly—but with precision, with perfection. To an adolescent, perfection is a seductive idea. When you can't do anything right—not family, not school, certainly not girls—the possibility of doing one thing well has enormous appeal. Since missing a dive hurts, the sport even elicits a degree of focus, at a time when focus on anything other than certain organs and patches of skin is problematical. Besides, you got to jump as high as you could, fly through the air, fall into water, and swim out again. Sometimes people even watched. What better way for a kid to fill time between shifts as a lifeguard?

That was also when I contracted my first case of hero worship. I had to learn to dive from someone, and Tommy Ortiz was the best around. He was a few years older, short, dark, compact, and genuinely gifted. He was something of a gross-motor genius, in fact, with a startling natural grace and talent at everything from Ping-Pong to football but a significant inability to keep himself scholastically eligible for any of them. On the diving board he was an explosive vertical block of granite, getting breathtaking lift out of the stiff wooden springboards of that era; in the air he was simple precision: straight up, straight down, no splash, whatever maneuvers you care to name along the way. Every time. I watched him so hard I knew exactly how he would do every part of every dive, and then he would do it exactly that way. Why did he keep practicing, I wondered, overlooking my own idiotic pleasure in doing the same dive over and over again, however badly. Tommy and I began spending far too many hours practicing. I believe we thought of ourselves, whether we admitted it or not, as the entertainment.

Eventually I would find swimming to be a subtler but more fascinating form of physics than diving. You press on the water to make yourself move through it. Learning to press on it with all the various planes and surfaces of the body, within the limited ranges of motion available, is a complicated task. An understanding of the physics is reassuring but not much help: you swim entirely by the feel of water on your body. The legendary swimmers—Weismuller, Spitz, Biondi—have been given not only remarkable physical strength and coordination but a kinesthetic sense that puts them in a different universe from you and me. They get more information from the water and respond to it with more precision, not to mention power, than the rest of us will ever dream.

The idea of pressing on the water with various body parts, when soft, cool water feels so good to those body parts, does bespeak a certain titillation, I suppose. No surprise, then, that along about late high school, swimming began to take on a deeper coloration. I'd been having so much fun that I hadn't given much thought to the fact that not only did one do it as nearly naked as possible, but girls liked to do it too. Once this came to my attention, whole new areas of exploration emerged.

Back to the Dittlinger dam, for instance. One spillway made a lovely waterfall, tumbling over a stone staircase. Slide carefully sideways along the middle step and you could sit under the waterfall, breathing comfortably, totally hidden. Few knew this. There was room for two. Once there, concealed, you could do pretty much what you wanted to, even remove clothing. Right out in the Texas out-of-doors at high noon of hot summer days. Get overenthusiastic, though—raise your head or other body part too high—and the force of the water peeled you off and hurled you, in whatever state you happened to be, into the pool below. This represented a delicious challenge.

I immediately fell in love with the longest pair of legs in town, the local Esther Williams, to whom I began playing a kind of pimple-faced Weismuller. Esther was as swim-crazed as I was but a couple of years older. She organized a series of amateur water pageants, mostly to indulge her passion for what was just becoming known as synchronized swimming. To fill time during costume changes, I led a troupe of clown divers: funny bathing suits, rodeo jokes, watery pratfalls. Let's hear it for human dignity.

With a few dry hours a week still to kill, I gave racing a try. Our high school offered no scholastic competition (or pool), but there were state championships. I won bronze medals, discovered how much lactic acid hurts, and, Esther being off at college, fell in love for five minutes with a female racer. I threw in for a time with the racing crowd and buddied up with another diver, call him Dan Shaw, who was perhaps not quite as talented as Ortiz but was definitely funnier. Once I saw him reach out of his car window and swipe a glass of iced tea, as he drove by, from the tray that a jaywalking waiter was carrying across the street. That kind of guy.

I thought swimming might pay for college, but Texas A&M was the only school that showed any interest. They admitted me to the athletic dormitory and allowed me to mop dressing rooms for thirty-five cents an hour. A few weeks into the fall semester, who should walk in off the street but Shaw, asking for a tryout. One look, full scholarship. I was glad to have a friend around. Two weeks later, after dazzling us at every workout, Shaw announced he'd gotten a better deal—playing quarterback at another school—and disappeared. The coach told me later that he'd been unable to produce a high school transcript. That winter Shaw and some other buddy pulled a very bad bank robbery, were immediately caught, and went to the slammer. It occurred to me that only a kind of twisted luck kept me from being an accomplice. I never heard of him again.

With or without Shaw I was miserable in Aggieland—fatally immature, homesick, pining for Esther. Loved the swimming, hated everything else. In *Haunts of the Black Masseur*, a strange and wonderful examination of obsession with swimming, author Charles Sprawson speaks of the swimmer as "someone rather remote and divorced from everyday life, devoted to a mode of exercise where most of the body remains submerged and self-absorbed. It seemed to me that it appealed to the introverted and eccentric, individualists involved in a mental world of their own." He certainly had my number. I spent the Christmas break swimming at the dam alone, brooding. When the semester ended in January I turned in my mop and transferred to North Texas State, which didn't even have an indoor pool.

After one nonswimming semester—the spring one, to make things even squirmier—I was seized by a whirling hyperactivity. The real-life Esther Williams was making sarong movies at the time, and it was obvious that my own hometown Esther and I were meant to live in bliss on a tropical island somewhere. I offered my racing services to the University of Hawaii, to a shocking silence. Esther and I had a terrible fight by mail. If I would just give up this English major nonsense and become a high school football coach, we could get married and start having babies. I was nineteen at the time.

The next fall, after the usual summer as swimming pool cowboy, I returned to North Texas State early to get first crack at part-time jobs. Waiting for classes to start, I decided that Oklahoma U. wouldn't make Hawaii's mistake, hitchhiked to Norman, found the coach, and explained my plan.

No thanks. Don't you even want to see me swim? I asked, remembering Dan Shaw's one-workout triumph. Nope. I hitched back to Denton, forswore swimming forever, broke up with Esther, married someone else in December, and started having babies. By then I'd turned twenty.

This is not to imply that water was always a benign and accepting medium. My other near drowning, more than a decade after the first, was at the dam. Envision a stone-walled cubicle with fifteen-foot walls open to the sky. Water thunders in by waterfall at one end and out the other through underwater flumes. The water in the room is ten feet deep, with three feet of foam on top of that. To enter you dive from a wall or shoot the falls; to exit you swim out underwater. Along each wall, a few inches below the surface of the water, is a three-inch ledge on which it is just possible to pull up and stand—if you succeed in catching it while you still have air. If you miss, you can't get your head above the foam to grab another breath. Catch the ledge, or submerge and swim out.

I was perched on the ledge one day when a stranger about my age appeared on the wall above. What I was doing must have looked like fun, because he jumped in. When he didn't reappear I assumed he'd swum out underwater. Then a hand rose above the foam, I reached out to help, and was immediately snatched off the ledge and into a wrestling match with a strong young man in desperate trouble.

I tried to grab the ledge again—and again, and again, as we were swept around in the eddy current—but kept missing. He was on me piggyback style, and the Red Cross rescue moves weren't working. Running out of air, I was trying to estimate if I could drag him with me down and out the exit—a long underwater swim with a panicky passenger—when the ledge brushed by one more time. I managed to grab it and pull us both up into the sweetest air either of us had ever breathed.

We braced in a corner for a few moments, gulping, congratulating ourselves for surviving. Shouting over the roar of the falls, I explained to him how one exited the chamber; he explained to me that he couldn't swim. His friends lowered a rope, and he climbed out. I still can't understand how a nonswimmer could jump into so violent a piece of water without even knowing how deep it was. I felt very strongly afterward that it had not been strength or skill that had saved me, or some Nietzschean will to live; it was simply the oxygen-using nature of my cells that caused me to wriggle my way out of the fix I was in. It was almost nothing that I had anything to do with.

I'm mystified now that these near misses didn't frighten me away from swimming. The physics—working the water with the body, flinging the body through the air—seems to have been more important than the biology, even if the biology was my own precious self. The joy of working the water seems to have taken me out of that self—which, as a teenager who tended toward surliness when I wasn't wet, I needed perhaps more than I realized.

Swimming at the dam wasn't all idiotic dangers, of course. Most of my time there was spent in some kind of courtship, splashing in the waterfalls, lolling in the sun. But I'm glad I had the other, nervy part too, the outlaw swimming. I don't defend or advise it, but I count the experience useful. Every other quarry rat, every urban wharf diver, every kid I ever knew who found a natural body of water to be the appropriate setting for spinning off excess energy, tells me the same thing. Water was one of the best parts of our growing up, and the best water we could get into—which was usually also the most dangerous—remains in our minds forever as a very precious place. We could never quite believe that the grown-ups weren't suddenly going to turn up, run us off, and close it permanently against our use.

6

dark woods, dark waters

In 1903 an explorer named Leonidas Hubbard, deep in the Labrador bush, ran out of supplies and, even though he continued to catch and eat a fair number of trout, wasted away on the lean fish alone and died. Aboriginal people have always understood this danger, preferring to settle where game is not just plentiful but fat—which means where there is plenty of water. Water makes sap, sap makes sugar, sugar makes fat; fat keeps you alive. An abundance of rainfall makes for lush vegetation, which makes its way through the food chain and helps make everything else lush too. When in doubt, head for water.

Of course water is even more important in its pre-fat form. Explorer Vilhjalmur Stefansson pointed out that running out of fuel in the Arctic means dying not of hunger but thirst, with no way to melt the water that is all around you. Stefansson knew about fat too. By hard experience he became something of a dietary nut, settling finally on what he considered to be the perfect ratio: 20 percent lean meat, 80 percent animal fat. He particularly loved butter. He lived to be eighty-three.

The wilder woods in the Northeast have plenty of water in them, the most enjoyable of it in tumbling rivers and clear or dark-water ponds and lakes, but access can be difficult. The banks and shorelines of New Hampshire and Vermont are heavily settled, and Maine's extensive forest is a long trip from just about anywhere. The greatest aggregation of readily available recreational freshwater in the Northeast is in New York's Adirondacks.

I had lived in New England for nearly thirty years before I understood what "Adirondack" meant. All I knew was that it was some kind of big park up in the mountains north of Albany. I've run across a lot of Easterners afflicted with the same haziness. If you didn't go to camp there, or have a family cottage, the Adirondacks tend to be just a name in the distance. Part of the confusion is the name. The Adirondack Park is in the Adirondack Mountains, but "Adirondacks," plural, is generically used for the entire region. Getting a fix on it can be difficult for the casual visitor.

Even if you penetrate the interior, you get the feeling you're never seeing much of it at a time. For high country it has a strange way of closing you in. The Adirondacks can be a gloomy place: drooping conifers, tannin-dark water, scruffy winter dieback, stark rock amid lowering skies. At times the combination of rough country and atrocious weather gives a strong sense of foreboding. One view of their history makes gloom seem entirely justified. The Adirondacks have always had a way of bringing a sharp discipline to inappropriate users.

Samuel de Champlain might be said to have started it all. In 1609, only a few days after he graciously donated his name to the lake that marks the Adirondacks' eastern boundary, he potted two Iroquois with his trusty harquebus, touching off Iroquoian detestation of the French that lasted until, and materially contributed to, France's eventual eviction from the New World. Champlain was followed by a Black Robe, the Jesuit Isaac Jogues, who was very likely the first white to penetrate the interior of the Adirondacks—albeit unwillingly, dragged there as captive of the Iroquois in 1642. After they bit off some of his fingers, tore out the remaining nails, and otherwise editorialized on the subject of the True Cross, he escaped; when he returned a few years later, having failed either to deliver his message or to comprehend theirs, they bashed his head in.

It was an inhospitable place, one the Iroquois themselves didn't much like. When wandering Algonquins attempted to winter there, and nearly starved, the Iroquois called them, smirkingly, "bark-eaters"—which in Iroquois is alleged to sound something like "Adirondacks." Until the American Revolution the Adirondack wilderness functioned principally as an escape route for Algonquin-French raiding parties preying on settlements to the east. No prudent pursuer would follow anyone into the interior of *that* country.

The first time Chris and I went there was shortly after Labor Day, when, although we didn't know it at the time, the Adirondacks are essentially closed. We were hungry by the time we arrived in Long Lake, and pulled up in front of a saloonish-looking restaurant that appeared to be open but was not. The sign on the door said NO GUNS OR KNIVES IN RESTAURANT.

I may have given the impression that we found the Adirondacks while searching for canoeable rivers, but in fact it was the other way around. We hadn't yet taken up canoeing and had just driven over for a weekend look, with no particular destination in mind. We'd made no reservations but found a funky old family resort in the hamlet of Blue Mountain Lake that would put us up. There were canoes for the use of the guests, and we thought we might give that a try the next day, if the nasty north wind had settled and the lake calmed down a bit.

Instead, dawn brought a not untypical chilly fall rain, which drove us indoors to the Adirondack Museum—home of, among other things, a fine collection of watercraft. The classic Adirondack guideboat is featured, along with a selection of other lovingly restored boats, including some historic canoes. One of them had Chris circling back for a second look, and then a third. It was the *Sairy Gamp*, a tiny cedar slip only nine feet long, in which an outdoor writer named George Washington Sears, pen name Nessmuk, had toured the region in 1883.

We strolled through the other exhibits, went back out into the rain, and drove home, essentially giving up on the Adirondacks. The *Sairy Gamp*, however, had lodged almost obsessively in Chris's mind, and a few weeks later she went back. She spent some time in the museum's library doing research, loaded up with source material, and came home with an idea. Sears had taken his tiny canoe from the southern to the northern edge of the Adirondacks and back again, a trip of nearly two hundred miles. Chris wondered if that trip could be duplicated in modern times. What if, more than a hundred years later, we gave it a try? It was a shocking proposition. Neither of us had ever done anything like that before. Chris had never spent a night outdoors. But the trip seized our imaginations, and we spent the winter making plans.

The next spring, proud new owners of two tiny solo canoes, we began looking for a shakedown cruise. A local guide recommended a portion of Sears's route that was easily accessible, undemanding, and, with a one-and-a-third-mile portage, would give us a well-rounded three-day introduction to the work ahead of us. That's how we discovered not only the Raquette River in springtime but, for the rest of our lives, the most enjoyable way we've yet discovered for getting to the most beautiful waters we can find.

Although Adirondack history doesn't really get rolling until after the Revolutionary War, two prerevolutionary entrepreneurs did try to settle there, and inadvertently set the tone of the place. Philip Skene, a Scot, sided with the British and lost everything; his competitor William Gilliland, an Irishman, backed the rebels—so the British burned *him* out, and he went to debtor's prison. Mulling his losses after he was released, Gilliland went for a walk in the winter woods and established something of an Adirondack tradition by freezing to death.

The string of would-be developers who followed the revolution continued this tradition, going bust or meeting with other misadventure with a regularity that would be hilarious if it were not so grim. There were the two John Browns, for example, in Adirondack history. The less well known was the Rhode Island merchant who founded Brown University. In

1798 he acquired the huge parcel of Adirondack real estate still known as the Brown Tract, and divided it into seven townships, named Enterprise, Perseverance, Unanimity, Frugality, Sobriety, Economy, and Regularity. Settlers turned up their noses. (Perhaps a sunnier salesmanship was called for. Frivolity? Inebriety?) When Brown died, a son-in-law, Charles Frederick Herreshoff, managed to clear some of the immense acreage, build a manor house, and even have a herd of sheep driven all the way from Rhode Island. When none of these labors proved profitable, he switched to mining and managed to produce a total of one ton of iron before he went broke and shot himself. (Adirondack luck set in early: one of the few settlers who did choose Brown's Tract died, in 1817, from being kicked in the testicles by an ox.)

Herreshoff Manor was then taken over by Nathaniel Foster, who settled a feud with a local Mohawk named Drid by gunshot and, fearing revenge from Drid's relatives, fled to Boonville—a town named for Gerrit Boon, who went broke pursuing a gigantic maple-sugaring scheme. (The wooden troughs Boon misguidedly used to collect the sap warped into uselessness in the winter cold.) Herreshoff Manor next became a hotel, run by one Otis Arnold, who killed a local guide in a dispute over a dog collar and, in despondency, filled his pockets with stones, rowed out into a nearby pond, and rolled over the side.

The other John Brown, the abolitionist, was trundled home to his Adirondack farm—after his hanging—and thus fetched up there permanently, buried in North Elba. The founder of the Adirondacks' most successful iron works, David Henderson, died when his pistol went off in his knapsack. William H. H. "Adirondack" Murray, the minister who in 1869 published the book that started the region's first tourist boom, lost wife and pulpit over matters of character, inattention to duty, and race horses.

The most celebrated Adirondack promoter of all was W. W. Durant, who invented the Great Camps that brought the rich New Yorkers—Morgans, Vanderbilts, Whitneys, Rockefellers—to the Adirondacks. Durant founded the town of Raquette Lake, built a railroad, and acquired holdings worth an estimated $1.2 million. His backer had the bad timing to die at a moment when Durant was grievously overextended, however, and he went predictably bankrupt, lost everything, and ended up clerking in a hotel in the tiny town of North Creek. (The Great Camps still standing are now white elephants, periodically sold for taxes or given away, charitably, to institutions that must then struggle to afford the maintenance.)

So it went. Alvah Dunning, one of the most colorful of the fabled Adirondack guides, died in Utica in 1902, either because the gas light in his hotel room was faulty, or, as one story has it, he blew it out instead of turning it off. Pulp writer and noted scoundrel Ned Buntline didn't come to the dreadful end he deserved—he survived an attempted hanging—but his teenaged wife died in childbirth on Eagle Lake, and he had to hide the next wife's shoes to keep her from running away. Chester Gillette died in the electric chair for drowning his pregnant lover, Grace Brown, in Big Moose Lake in 1906, providing Theodore Dreiser with the plot for *An American Tragedy* (and George Stevens with *A Place in the Sun*, a film with Montgomery Clift and Elizabeth Taylor). Verplanck Colvin, who first surveyed the Adirondacks and almost single-handedly convinced the state that the wilderness must be protected, ended up muttering his way about the streets of Albany, and died alone. Bob Marshall, whose youthful love of the Adirondacks impelled him to found the Wilderness Society, died in his sleep at age thirty-nine. "Adirondack Harry" Radford, who personally organized efforts to reestablish both moose and beaver, was stabbed to death by a native guide on Bathurst Inlet on the Arctic Ocean when he was in his early thirties.

If a region of the earth can actually be charged with malevolence, the argument might be made that the Adirondacks had justification: while the wilderness was writing off all of these people, others were raping it for its lumber (and a short-lived but hideously polluting tanning industry), burning it to cinders in serial forest fires, despoiling it for its minerals, and hunting and fishing it within an inch of its life. We managed to extirpate its moose, eagle, wolf, lynx, panther, and most of its beaver in less than a hundred years, although beaver have been thoroughly reestablished and eagles and moose are making a tentative comeback. The volume of flow in its rivers has decreased steadily since the advent of lumbering. (The professed purpose of setting aside the park's huge acreage was to protect the Hudson River watershed, not to provide recreation for rich or poor.) White people came to this enormous wilderness late—we knew the source of the Columbia before we knew the source of the Hudson—and didn't even call it the Adirondacks until 1837, but forty years later it was "hunted out, timbered out, overrun . . . and ruined," according to William Chapman White, in *Adirondack Country*.

Or maybe the string of human disasters was only a psychological response to the miserable climate, the pestiferous insects, the thin and acid soil of the place. When the public finally began to arrive in significant numbers, a considerable portion of them were tuberculars, driven to desperate measures. Medical science of that day believed that wilderness air restored damaged lungs.

What most people mean by "Adirondacks" is the High Peaks area, the mountain range rather than the whole Adirondack Park. The park is such an amorphous mass, such a hodgepodge of public and private lands, that only its residents seem to have a clear idea of what actually constitutes it. It's what's inside the Blue Line, that's what it is—the infamous line drawn on the map, almost arbitrarily, to establish the park's original boundaries. And throughout its history it has been fought over, and is still being fought over, as rancorously as any ostensibly "preserved" piece of real estate in the country. Back in 1894 the legislature declared it, famously, "Forever Wild," and has reconfirmed that status from time to time, but has left the details of its wildness for others to resolve. Resolution keeps escaping; democracy keeps bumping up against elitism. As historian and sympathetic preservationist White puts it, "the woods locked up against the few meant the woods opened up for the many." What has resulted is an extremely nasty hundred-years' war between preservationists and developers. The population tends to be fiercely loyal to the Adirondacks, fiercely historic minded, and fiercely either pro- or antipreservationist. Peace is nowhere at hand.

However politically amorphous it may be, the park—the six million acres inside the Blue Line—is geographically coherent, a single high plateau that geology and erosion have divided into five watersheds. These flow, respectively, into Lake Champlain, the Upper Hudson River, the Mohawk River, the Black River, and the St. Lawrence. The Black and

Mohawk Basins are mostly hunting, fishing, and agricultural country, not terribly accessible, not terribly attractive to the short-term visitor. The Lake Champlain Basin contains most of the High Peaks; the Hudson Basin suffers from the most development; the St. Lawrence Basin holds the best canoe country. (The Raquette River is in the St. Lawrence watershed.) Paul Jamieson's *Adirondack Canoe Waters: North Flow* is the best guide I know to the Adirondacks—and, incidentally, worth reading for the sheer grace of the writing. I've drawn on it heavily throughout these pages.

Actually, our *Sairy Gamp* visit was not the first view of the Adirondacks for me. I'd been to Lake Placid back in the early seventies, on assignment for *Skiing* magazine. Lake Placid had hosted the Winter Olympics in 1932, and visits to Cortina and Squaw Valley had convinced me that holding a Winter Olympics meant a kind of doom for a ski resort. Lake Placid, then a flyblown little town with ancient and rapidly deteriorating facilities for the various winter sports of forty years before, confirmed my notion. Holding an Olympics, summer or winter version, requires a great deal of capital improvement. Bond issues and public subscription pay for bricks and mortar but not for upkeep.

Whiteface, the ski mountain a few miles north of Lake Placid, was not much more appealing. State owned and run, its revenues came only from providing basic service for skiers, rather than the condominiums and second homes that were then bringing boom times to competitors in Vermont and New Hampshire. (Whiteface's rampant unprofitability never stopped privately owned ski resorts elsewhere in the Northeast from complaining about unfair competition from "socialized" skiing.) Lifts and trails were ill maintained, the lodge was as unappealing as a Russian warehouse, there was a kind of industrial feel to the whole place. Its only selling point was an unusually large amount of vertical footage, which skiers treasure because in the right conditions it allows nice long runs. On Whiteface, however, the upper third of the mountain was in those days hard-core experts' terrain, unskiable for most of the population for most of the season. I hated the place.

Many millions were poured into Whiteface for the 1980 Winter Olympics, and it is undoubtedly much improved. I haven't been back to ski, but I have seen the town of Lake Placid, which is now painfully overbuilt and, following the example of nearby Lake George, seems to be trying to become a kind of high-country Coney Island. None of this is surprising, since death, in the form of a Winter Olympics, has kissed Lake Placid twice now.

At least Lake Placid has had an infusion of prosperity; the rest of the Adirondacks can seem horribly depressed—and sourly angry about the whole thing, for which there's also considerable justification. Visitors by automobile are particularly frustrated. Finding something to eat other than franchise food can be problematic. You often find yourself driving long distances down aisles of scrubby trees interrupted only by an occasional roadside bog. You can drive alongside some of the lakes, at least for short distances, but their best views, like those of the surrounding mountains, are usually inaccessible.

The town of Old Forge has joined Lake Placid and Lake George in pursuit of rampant tackiness. Raquette Lake, once a thriving railhead, is now a ghost town. Blue Mountain Lake, smack on one of the most beautiful bodies of water in the park, is a string of small marinas and outfitters, a restaurant, a gift shop, and its magnificent (and wildly popular) museum. Long Lake actually offers less, and Tupper Lake, which used to be the Adirondacks' industrial center (toothpicks and wooden ice-cream spoons), looks as if it has gone permanently broke. The most popular style of Adirondack architecture might be called Boarded Up. The fact is that you can't get *at* the Adirondacks by car. Yet if the Adirondacks have been ruined, as every old-timer will belligerently assure you, it was the automobile that did it, killing off the fashionable exclusivity that made the "Great Northern Wilderness," as it used to be called, the dream destination of the crowded East Coast. That's all over now, of course, as with Yogi Berra's ex-favorite restaurant: nobody goes there anymore, it's too crowded.

The reader might at this point expect me to turn this around and begin explaining how terrific the Adirondacks really are. Well, let's see. When the weather is good, the Adirondacks aren't gloomy at all. That's about one day out of four. The rest of the time you're likely to be experimenting with the wilder shores of hypothermia—although summer temperatures of well over a hundred degrees have been recorded. To avoid the crowds you have to go in September, and September is when it rains. They say October is better, but it's so cold in October that only the daring get to see the place.

We've been driven out by blackflies on May 5 and by snow on May 13; we've also been kept indoors by snow on September 20. The earliest post-blackfly date I've attempted to go there is June 24; that time it was mosquitoes that forced me to flee, but the blackflies weren't entirely gone yet, and they were adequately backstopped by punkies and deerflies. National Guard fighter jets find the vast unsettled distances of the Adirondacks an appropriate place for nerve-shattering low-altitude practice runs. We've also regularly been baffled by mysterious distant rumbles that sound like thunder even on the rare clear-sky days. We think it's artillery practice at Fort Drum—a military reservation only five miles from the park's western border—but we're not sure. We hear it on almost every visit. Locals say they don't know what we're talking about.

On our *Sairy Gamp* trip, which took us twenty-eight days, we became irritated enough with motorboat traffic that at one point we abandoned Sears's route to steal a restful day in the St. Regis Canoe Area, where no powerboats—or aircraft—are allowed. Twenty minutes after we made camp, the state's own Department of Environmental Conservation helicopter came over, deafeningly, at about two hundred feet. This one wasn't shitting fish, it was on some other errand; twenty minutes later it returned, and twenty minutes later repeated its intrusion, and so on, flying some kind of shuttle course back and forth for the next three or four hours. Some damned wilderness.

We do go canoeing there every chance we get, I admit, and are sort of halfway looking for a cottage, but I couldn't say we recommend the place. There's nothing over there but trees and mountains, lakes and rivers and ponds, and the wildlife—what's left of it—that goes with all that. You probably wouldn't like it.

On our first trip there Chris found herself simply wanting to flee: the place spoke to her in some way that she didn't really understand, but she couldn't bear being there just as a tourist.

We had lived in similarly stark northern New Hampshire for twelve years, in a tiny and exceedingly rural mountain hamlet, and she'd formed some firm friendships there. The Adirondacks reminded us both of that time, but that wasn't quite what intrigued her so.

It took a few more visits for her to figure out the connection. Her parents had taken her to Canada's Laurentians as a child, and she fell wildly in love. The Adirondacks are a geological extension of the Laurentians, and the physical geography, the sensory experience, is virtually identical. Her parents exposed her to that world, she responded to it naturally enough as a child, and she's never stopped responding to it. In a sense she'd gotten from the Laurentians exactly what I did from the Kiamichis.

So what are the Adirondacks? Two thousand lakes, six thousand miles of rivers and streams, forty-three mountain peaks over four thousand feet. All of it in the form of a roughly circular granite batholith larger than the state of New Hampshire. We haven't experienced all two thousand lakes, only ten or fifteen of them, and a few of the rivers. We think about the place quite a bit, though. It's on the edge of our local daily weather maps, and we almost always look to see what's going on over there. Usually the Adirondacks are represented by an oval of dark clouds. But in spring, anyway, that means there's hardly anybody there.

7

the show

> The distant river-reach seen in the north from the Lincoln Hill, high in the horizon, like the ocean stream flowing round Homer's shield, the rippling waves reflecting the light, is unlike the same seen near at hand.
>
> —Thoreau's Journal, May 1, 1854

On a lake at sunset a backlit mallard paddles by, etching a vee-shaped wake on the mirrored surface. Each arm of the vee is uniformly chevroned, herringboned with ripples. The force of each foot stroke by the duck pushes up a small bow wave that marches back along the vee, reflecting pure silver on the top of its curvature, pitch black on the downslopes. As the duck passes, the wavelets diminish and diminish and finally are reabsorbed, first the chevrons disappearing and then the vee-shaped wake itself. Three otters play just offshore, dark round heads popping up in patches of silver reflection. One holds a small fish he keeps showing to the others, nyaah nyaah. A tiny point of white blinks in the distance, reflection off the wet white breast of a loon. Loon dives, point of white disappears, ripples spread. Water and light.

We wake to the usual gray haze, as if the silvery reflected water surface of the dusk before has been transmuted overnight into morning fog. It lifts; the morning's paddle is in dead-calm water along a shoreline of sheer rock fifteen feet high, reflected perfectly in the mirror-smooth water over which it looms. The rock surface that curves back away from the canoe above the water appears to do the same below the waterline, and on the lower side the reflection disappears into the depths, into the upside-down version of the woods above, an unattainable magic kingdom: unimaginable depths I long to dive down into.

Halting for lunch, sprawling on smooth granite, we inspect the sudden outbreak of cumulonimbus, curling at the edges and shouldering lacy ice caps above them, demonstrating such boiling energy that they make me remember all that time I spent looking at clouds as a kid, watching thunderstorms erect themselves in the sky over Texas: physics made visible.

As you near the end of a portage you begin looking above the trail for a patch of light, a glimpse of sky. An opening tells you that water is just ahead: you'll soon be able to put down the load you're carrying. The hole in the forest canopy signifies relief.

Lakes and rivers make large interruptions in the canopy, but smaller holes also let in light, and demonstrate its effects in surprising ways. We live on a wooded hillside laced with old logging roads and connecting trails. One fall a storm blew the top out of a tall white pine beside a trail, toppling the wreckage to the floor alongside the trunk. The hole it left allowed it to be dramatically spotlighted, as if with atrium lighting, drawing enough attention to itself that it became, for a while, a significant landmark.

Later we'd notice not so much the light let in as the scent from dying needles, aromatic pine oils baking out in the sun. At first I thought of this as the result of heat rather than light, but light is what it is, radiant energy, only a different part of the spectrum. It is strange to think how narrow our spectra are, how limited the conditions that fit our sensory capabilities. When the sunset fades from gold and magenta glory to slate gray coldness, nothing has changed but a few degrees of temperature and the angle of the light, but we, so restricted in our receptors, consider that the beauty has faded.

Once I'd noticed that hole in the canopy, I started looking for others. There was another obvious one a hundred yards farther up the hill, and, spotlighted directly below it, a patch of rich grass—stimulated by the light—in an otherwise barren stretch of trail. Another bright spot, above an old logging road, has gathered moss on its center hump, and ferns march along the shoulders. Clearly, holes in the canopy attract their own small biological communities. A shaft of sunlight makes an edge, and edges bring changes.

Some of the gaps are caused by trees that have died and gone bare, but still stand. These "snags," a forester tells me, make veritable wildlife condominiums. I'm sure he's right, but maybe snags, like shafts of sunlight, are simply making more visible the normal density of life in the forest. Only in shafts of light do you see clearly the swirls of flying insects that are usually there in the woods. Holes in the canopy might make interesting research. One might measure, for example, ground temperatures or air temperatures within and without these shafts of light, these columns of slightly different air—different in content as well as quality. The shafts of light must be analogous to the thermals on which hawks and buzzards soar. There must be life forms that soar these columns too.

We can't see thermals, only see the hawks that ride them. There are surely other edges we can't perceive. We don't know what all of the edges are. What are edges to us may not be edges at all to other creatures, and vice versa. The towering thunderhead looks utterly solid, but its hard-edged outline marks only where air meets air: a minor shift in temperature and moisture, a change so subtle that only when conditions are right can we see it. On the clearest of blue-sky days, when there's only a bowl of emptiness to our narrow range of vision, there must be similar towers, tumbling masses, mare's tails and sheets and billows, clouds all over the sky, if we only had the sensors to pick them out.

On a breezy day I watch the column of heat over a campfire as it whips back and forth, a colorless flame made visible only by the distortion of the background seen through it. An Arctic canoeist tells of stepping onto a low ridge and looking back at his campmates below, each standing at the base of a twenty-foot column of mosquitoes: twenty-foot columns, in effect, of moisture and carbon dioxide, made visible only by the mass of live protein swirling in them.

A breath of breeze throws highlights across the surface of the water as if scattering coins; the reflecting facets tip-tip-tip from side to side, from front to back, silver patches that break up the black, black patches that break up the silver.

Wavelets tap-tap against the rock shore, against tree trunks lying in the water, an aural rhythm to go with, or sometimes play against, the winking light. The liveliness of water is in its erogenous rise and fall, ebb and flow. Wind stacks it, raises it, pushes it into motion; gravity pulls it back down again, quiets it. A floating reed in the pond makes a slash mark of gold on dark water. Wavelets of dark green march along its length in perfect Rockette rhythm, in what looks like utter purposefulness. The wind drops, the wavelets subside, the march halts. The sun drops lower, the golden slash mark turns into a green stick in muddy water. Light on water.

Art directors work with PMS colors, the acronym referring to the standardized Pantone Matching System. When a specific shade of ink is required from the printer, it is asked for by its PMS number. I was thinking of this while standing on a narrow spit of sand beside an Adirondack lake. There was a slough behind, the lake in front, and I was transfixed by the vertical 360 degrees of scenery I was standing in, wondering what the colors really were. The ground beneath my feet was sand; next to it were dark green reeds and lime green grass, not yet fully established. The mud at the edge of the slough was purple with a sheen of gray, beside black water. In the midst of silver-white reflections in the water was a blue-gray cedar stump. On the far bank of the slough the iridescent green grass was dazzling, the most vivid single color in the landscape. Taller grasses with yellow-green seed heads grew beyond that; above them were the green-black shadows thrown by various shoreline dendrology—white pines, maples, spruce, balsam fir.

The taller white pines were backlit with the evening's salmon-pink sunset; overhead, blue sky showed between strips of white cirrus, with clear blue showing to the east, opposite the sunset. In the eastern distance the late-evening strata were erasing themselves into a steely blue-gray, anticolor, noncolor, and on the absolute horizon, at the rim of my visual world, was a faint line of peach-colored towers, marking someone else's summer evening thunderstorms.

To the south there was a dark blue line of low hills, with a border of richest dark blue-green conifer forest below that. A sharp, pencil-thin moustache of yellow sand, then water, from the near horizon to my feet, all of it a slate-gray-and-brown, almost the definition of colorlessness but with a high component of silver-blue: the molten version of the dissolving overcast above. Something about this brownish tannic lake, chopped up by the breeze into restless silver facets, made the whole thing look purple.

Art director, would you please supply PMS numbers for the above?

Under bright sun the uninterrupted forest canopy itself is spectacular enough. Hardwood leaves are more or less translucent, and seen from underneath, with sun backlighting the top layer, turn an electric green, giving each tree a corona of light. Where the top layer isn't uniform, lower limbs reach out, tacking cell onto cell to fill in the blanks; where the sun reaches through, the lower layer glows too, making layers of blazing green stacked through the top twenty feet or so of canopy. The idea for the chandelier has an obvious source. Here and there no foliage interrupts the light until near ground level; then some seedling, lit by a spear of sunlight, becomes a waist-level burning bush, God's own table lamp: the enormous room of the forest, furnished with designer light fixtures. The sources of the stained-glass window—and the vaulted cathedral that it illuminates—are equally self-evident. We do great work with stone and glass, but the woods, using softer materials, do better.

Windbound on a brilliant hot day, I sit on a granite point, gazing at the incipient whitecaps in the distance. Langmuirs—those long lines of scattered foam that stream off downwind from no discernible source—march toward me. The water presents alternating planes of brown and blue, the translucent dark brown from tannin in the water, the deep blue from the reflected sky; brown plus blue equals purple, of course, they taught us that in third grade. The horizon is lumpy with hills, the wind-tossed lake surface lumpy with whitecaps, the sky above lumpy with cumulus clouds. In between, in the distance, is the perfectly flat waterline, its flatness duplicated in the browse line of white cedars along the shore. Lumpy below, lumpy above, flat in the middle, a flat line drawn around the circumference of the available visual universe. On a lake you are circumscribed.

No haze, a hard-edged, reality-based day: the shadow of the overturned canoe on the rock, of layers of the rock on itself, of my bare foot on the rock, are sharp, perfectly aligned, knife-edged: concretely located. Unambiguous; not subject to interpretation. Light putting the day under a microscope. Looking at light on water is the best way I know of enjoying the earth.

On another afternoon on the hillside behind my house, after a rain, slanting beams cut through the canopy, but the air has been washed so clean that there are no columns in it. The beams are there, I can tell where they are, but I can't see them, don't know *how* I can tell where they are. I walk carefully all the way around one of them, this apparent column of nothing, and can't see a thing—but there is a presence in front of me. I can see it only if I stick my hand into it, giving the light something on which to fall. A spot in the sky, a spot on the earth, nothing connecting them but a straight line: if the light beam doesn't hit anything it does not, visually, exist. Light goes out from the sun equally in all directions, but to our limited sensors does not exist except where it touches us or something we can see. As if the sun shines its light on Earth only because we are here to see it.

Or the other way around: perhaps our reason for being is for the sun to have something to shine on; to turn its energy into light, to become visible, which as far as we are concerned it does only on the surface of the moon and in the dusty atmosphere, on the lands and waters of Earth.

Holes in the canopy are the result of ice storms, blowdowns, disease, or, sometimes in our local woods, the rampant fox grape, which will envelop a tree and pull it down, sometimes taking neighbors with it. The mechanics that make the holes aren't always apparent, but you can usually figure them out. There's an exactitude to the transactions, almost a business of ramps and levers, wheels and gear teeth. We must have first understood the lever and the ramp from nature, and appropriated them. The individual who "invented" the wheel was surely someone who had a log roll out from underfoot, or a round rock. Fire was not invented but stolen, just as the old legends say. If we continue to pull away from the legends we will eventually take credit for inventing light itself. See our cities, our malls, our ginzas: we'll forget that we are given light, along with everything else.

Or, more accurately, it is the light itself that gives us all the rest. What the canopy finally makes me see is that light grows the forest. The woods are a product of light, as are the rivers and lakes and streams, sucked up out of the ocean by light, transported across distances by light's engine, dropped on the mountains and hills to form the watersheds by which our geography is finally organized. Light is the gift from which all of the other gifts come. All we have or are is only interaction between real estate and the electromagnetic spectrum. Light meets dust, and everything happens: glowing leaves, forest and field, blood, bone, martinis, bulldozers, Bach, trout.

Transfixed by a riveting view, I try to figure out why it is so beautiful. It is beautiful to me because it fits the frequencies and ranges of stimuli that my receptors will take in; because it suits my cultural aesthetics, my sense of balance and unity and wholeness. It is undoubtedly ugly to some other set of sensibilities. All of the woods and waters, all of the natural world is as beautiful as this, precisely as beautiful and precisely as ugly. Nature is absolutely beautiful and absolutely ugly, always, at the same time, in the same place. Sublimely ugly. Sublime.

8

going outdoors

The weatherman on the evening news points out that the previous night's sharp frost marked the end of the growing season. It is a surprising reality check from a TV weatherman: here is where we are in the year, here's what the planet is actually up to. Then he drifts on to more urgent matters, the morning commute, the scraping of windshields. No further consideration of how many thousand square miles of North American plant life had been flicked off, at least for the season, by that meteorological blip, that hinge point in the natural year.

Outdoors, which is where the growing season occurs, is an undependable place; why on earth would anyone want to spend substantial amounts of time there? Contrast is probably the simplest answer, at least for our largely indoor society.

One wouldn't expect telephone linemen or roofing contractors to feel the need as strongly as office workers. On the other hand, people whose daily lives put them out in the weather tend to have a much more realistic, not to say appreciative, grasp of the exigencies of the physical world. Dealing with it must come easier for them. That ought to free up attention for the finer points.

Or maybe it's a false question, the product of what anthropologists call indoor thinking. Indoors, like adolescence and urbanization, is a recent concept. Indoor thinking narrows the focus, making the outdoors of little consequence and yet at the same time larger, more mysterious, more threatening than it need be. TV weathermen, curiously enough, specialize in indoor thinking: their weather takes place on a map, in symbols. Outdoor thinking sniffs the air, covers the firewood.

For all my talk about craving exposure to water, I'm more indoorsman than not, most of the time a numbed-out captive of the media like everyone else living by grace of, and too close to, society's commercial engine. What's worse, the success or failure of the work I do depends largely on the reaction it will receive months or even years after I do it. This is a dislocating way to work, although hardly unusual. It is not exactly a reality-based way to spend one's days. It is because I am juggling the future during so many of my waking hours, I think, that going outdoors carries so much release. It happens right now, when you do it; when you're there you really are there.

For years now Chris and I have lived on the edge of one woods or another, enjoying both the peacefulness within and the occasional flurrying activity that flows out of them. That proximity was enough, we thought, but something kept us wanting to go deeper. Eventually we found a way. Now we keep going back as if someday the woods are going to crack open and show us why, although of course they never quite do.

As it happened we were preparing for another Adirondack spring trip when I first got word of Jud's illness. There was no need to rush to his side, but the prognosis was bleak. Our trip had been planned with another couple, Dick and Susan Todd, months before, and their schedule was tight. I wrestled with whether to make the trip under a pall or simply cancel. After talking it over with Chris, we decided to go ahead. The Todds didn't know Jud; I'd try to keep any preoccupation to myself.

The trip started in difficulty, with a missed rendezvous, rain, unseasonable cold. We got on the water late and were forced to grab a campsite immediately to avoid making dinner in darkness. The Todds slept badly in summer-weight gear, and the next day dawned gray and damp. We set out upriver again in still-blustery weather, trying to convince one another that the sky was brightening.

Which, actually, it was; eventually warm sun broke through, and the trip took a radical swerve toward bliss. We were on the Bog River Flow, which doesn't flow much—or, unlike the Raquette, swell with spring flood. The paddling was pleasant work, the river offering a succession of red sandbanks alternating with grassy verges and overhanging hemlocks and pines. Loons would bob up ahead and then behind. An osprey stood guard on a tall dead pine while its mate flew above; then the two joined forces, circling over their home base, not so much hunting as staking out their personal column of air. Twice we sneaked up on grazing deer.

Occasionally there would be no sandbank on the shore but a sand shoal underwater, reflecting a warm, reddish brown glow upward through the water onto our paddles, the undersides of our canoes, our grinning faces: two middle-aged couples, after a struggle or two, on a gentle canoe trip in not quite the right season.

When the usual midday headwind came up, we found a sand beach on a small island and lounged for a while, amusing ourselves with a new pair of binoculars, enjoying their capacity to leap, visually, through space. Hand the glasses to the next in line and hear the recipient mutter with delight as a vague and distant horizon, or some small flutter of life, reveals itself in more intimate detail. One did not have to walk or paddle over there to see it better. No expenditure of energy required. What a deal.

We didn't get as far as planned, but far enough, and, one direction being about as good as another, headed back downwind, where we found a pleasant hillside campsite in a beech grove. The campfire brought out wine and cheese, accompanied by a wood thrush quartet. Dinner was lemon chicken, cucumbers in wine vinegar, fresh fruit salad: could this be camping? Afterward we built the fire larger and settled in for a long discussion about nothing much at all. At one point I noticed Dick, who is also a writer, standing slightly apart, staring fixedly into the fire. I'd seen him earlier alone on the riverbank, seemingly lost in thought. "What are you working on?" I asked him.

"Nothing," he said quickly. "Not a thing. I don't seem to be doing . . . anything."

Exactly, I said, that's just it, isn't it? Nothing happening, nothing here but woods and water, a prescription for boredom: so why isn't it boring? I told the Todds, briefly, the news about my brother. I had so far been unable to think about his illness—or any other future topic, for that matter—but on some level I'd never really stopped thinking of him.

There was no chance of boredom when you went camping with Jud. What he practiced was not so much woodcraft as play in the dramatic sense: assigned roles to act out, almost as if we were following a script; lists of things to do. We'd load scout knapsacks on our bikes and head out of town, stopping usually at the first place where woods and water crossed the road. On the earliest trip I can remember, my job was to dig a cupboard into a streambank to hold our one can of Spam and one can of pork and beans, while Jud busily erected a sapling bridge over a rivulet that an eight-year-old—such as myself—could step across. The Boy Scout manual must have suggested these improvements; Jud, who would go on to become a poet and professor, was a student of the literature even then. That night a cold rain passed through, and, with no tent, we awoke soaking in our thin cotton sleeping bags. We dragged them under the nearby highway bridge, which held off the rain but formed a perfect tunnel for a stiff north wind. Jud built a huge fire, but the wind whipped the warmth away unless you got close enough to risk burns. It was the clichéd version of every bad camping trip I ever heard of. For years I assumed everyone had had one of those in his or her early experience. I was eager to go again.

He was teaching me, if sometimes by unfortunate example, the essentials: respect the woods, take care with fire, plan ahead, maintain an escape route—a methodology for solution to the physical problems of the "wild." Crude methodology, crude problems, but then subtler problems have a way of dissipating on you out there, which is perhaps the point.

Fifty years later, on our first chilly night out with the Todds, I'd been amused at my own finicky desire to get the fire started with one match, in the time-honored Boy Scout tradition. One of the things that delighted me most about camping out, when I came back to it after several decades, was the way it sanctions, even glorifies, parsimony. Bring only what you need, i.e., find out what, refreshingly, you can do without. It's corny stuff, camping. It was demonstrated to me first by Jud.

Getting your kit together can be an entertainment in itself. There's a certain pursuit of the ideal, of the delusional notion of having everything you need in the size and shape and quality that's exactly right for whatever you want to do. Since things never work out that way, there's the perfectionist pleasure of permanent readjustment. Equipment lists are romantic in themselves anyway. I always enjoy the outfitting chapters of expeditionary literature. Making those lists must be fun, particularly when you're going to have porters (or enlisted men) instead of carrying everything yourself. I loved that in the film *Out of Africa* our hero included in his safari gear a wind-up Victrola on which to play the dinnertime Mozart, along with tablecloths and wine glasses, not to mention what must have been several pounds of wine to pour into them. The British navy is famous for taking along on polar expeditions such things as French horns and printing presses. On the other hand, George Washington Sears limited his "trap" to rifle, fishing tackle, axe, knife, compass, frying pan, blanket, ditty bag, tent, one shirt, one pair of socks.

Thoreau's Journal, November 22, 1839: "Linnæus, setting out for Lapland, surveys his 'comb' and 'spare shirt,' 'leather breeches,' and 'gauze cap to keep off gnats,' with as much complacency as Buonaparte would a park of artillery to be used in the Russian Campaign. His eye is to take in fish, flower, and bird, quadruped and biped. The quiet bravery of the man is admirable."

On the Bog River a chorus of demented coyotes entertained us overnight, but all slept warmer, and we arose to the customary low morning fog, the world revealed by soft-focus lens. The river where we camped ran between parallel eskers, which put us between yin and yang hillsides, the shady south side still winter-bare, the sunny north bank showing the first haze of pink from maples and lime green from popple. Hobblebush and serviceberry were in bloom. Feeder streams tinkled dreamily on their way to the river. Our campsite was surrounded by polka-dotted understory, the tiny leaves of early foliage precisely parceling out the available light among themselves. The best part of woods in spring is the light: there is so little foliation, so little understory shade, and the sun, when it is out, comes on so wonderfully strong.

By the time we'd finished breakfast the mists had drifted away upstream, and we took coffee down to the riverbank. A waning three-quarter moon still loomed over the opposite hillside, a blue-gray rock hanging in the sky. We talked about how improbable that looked, about the precession of planets, Fibonacci sequences, fractals, and other lovely subjects none of us knew much about. I amused myself swinging my attention from the boulder in the sky, viewed through binoculars, to the tiny little florets and mosses among which we lay, viewed through my bifocals. British soldiers, I thought at the time; they were not in fact that particular lichen, I would find out later, but out there on the riverbank the wonderful datedness of the name flooded over me. It was surely pre-Thoreau, a coinage of the revolutionary period, now putting our imaginations back in another time as effectively as the binoculars put us, momentarily, on the moon. Thoreau was a master of that imaginative leap. That Homer had heard the same mosquito hum, felt the same breeze on his cheek, was the principle in nature that moved him most.

Time was on my mind, I suppose. It has been proposed that eskers mark the tracks of prehistoric rivers, built out of the debris deposited as running water cut its way down through the ice cap. The peak of the ridge thus marks the former river bottom, making a height out of what once was a low, just as Isaiah promises. To be between parallel eskers is to be down a deep cleft in time, where canyon ice walls once stacked a thousand feet above. No wonder these woods tend toward the chilly.

Chris, who is Canadian, has turned me into an armchair Arctic explorer, plundering the literature if not the landscape. In one tale the expedition, well back in the bush, runs into a group of Inuit and attempts to communicate. None of the available spoken languages being of much use, both parties begin inventing sign language on the spot. One of the Inuits, attempting to explain something about hunting plans, begins waving his hands about and slapping at his neck and face. Finally the visitor is able to figure out what he's driving at: hunting plans for next spring, when the ferocious Arctic mosquitoes will again be out in force. Sign language does have a way of cutting to the physical realities.

The expeditioning I grew up with used equipment far less functional than the gear available today, and we used it to go into country that seemed far less benign than the woods and waters accessible to me now. Not that that made it heroic. Once we got past the hitchhiking days we usually camped out of the back of a car (or the front of a VW Beetle). From Texas we would light out for the Rockies just to get cool. I did more than one fifteen-hundred-mile circle, with various traveling companions, up through all the snow-covered mountains we could reach, sleeping on a tarp on bare ground without a tent, curling up in the car when it rained, miserable most of the time with fatigue and grime. We'd stop at an occasional Y for a shower, and at an occasional diner for a more solid meal than the hot dogs and canned goods we ordinarily warmed at roadside. We'd camp in a ditch if that was the best place we could find, and it often was.

We did it that way because it was the cheapest way to travel, and we wanted to travel, just to see things (pulling off dutifully for every scenic overview and historic plaque, looking, driving on, never two hundred yards from a road): traveling just to *go*. It would never have occurred to us to go outdoors because that was where we wanted to be. It never occurred to us that there might be inoffensive, accepting outdoor places available. We didn't know how to look. Going outdoors was the price you paid to go at all.

Unfortunately, if you can get there by car, it won't be far enough in and there will be too many others around, too close. This is undemocratic of me, I suppose, but there it is. You can't do it by car. If you drive up to the edge of the

Grand Canyon and get out and look, you'll see an awesome and beautiful sight, but it won't be the Grand Canyon, it'll be the Grand Canyon by car, a kind of virtual version.

Access is the crux. The desire to set aside and the desire to make accessible are mutually destructive. You can always set aside and protect, you can't always go back and remove accessibility. It is a conundrum. It was the mechanization of lumbering that made the north woods so accessible, and the mechanization of recreation that produced the RVs and four-wheel drives and other gadgetry that deepen that access, and—including please the all-terrain vehicles and water scooters—that so diminish and degrade the very thing we're given access to. (Imagine if you will the difference between the lake that is experienced by a jet-ski rider and the same lake as experienced by a canoeist.) It is mechanization of lumbering that gives north woods populations sufficient income and leisure time to enjoy the woods that the lumbering hasn't yet destroyed, that affords the recreational technology that allows the population to go deeper and deeper into the woods, to live deeper in the woods. When there are less and less woods to go into.

I come down firmly on both sides of this question. Our eighteen-foot canoe, made of high-tech Kevlar, weighs thirty-nine pounds. This allows us to go farther and faster and get deeper in, at a rather advanced age for this kind of foolishness, than we could ever manage with an old eighty-pound wood-and-canvas classic. Technology not only extends our range, it extends our time, giving us more years. It extends everyone else's too, however, and sometimes Chris and I fear we'll be crowded out—interesting locution—before we lose the capacity to go. Which is also elitist of me, I suppose.

• • •

". . . But it is not only to the real life of birds and flowers that the little rivers introduce you," writes Henry Van Dyke in *Little Rivers*. "They lead you often into familiarity with human nature in undress, rejoicing in the liberty of old clothes, or of none at all. People do not mince along the banks of streams in patent-leather shoes or crepitating silks. Corduroy and homespun and flannel are the stuffs that suit this region; and the frequenters of these paths go their natural gaits, in calf-skin or rubber boots, or bare-footed. The girdle of conventionality is laid aside, and the skirts rise with the spirits."

With the right gear, all you need is patience. If discomfort comes it'll be from bugs or weather; bugs can be impossible, but the weather will usually get better if you just wait it out. A lot of patience, sometimes: once on a lake in Maine, Chris and I were rained on, hard, for sixteen straight hours. We were comfortable enough and mostly dry in our tent for much of the time, but it was sixteen hours nonetheless. It was a little late in the season, on a September third, a date that is easy to remember because it happened to be our twenty-fifth wedding anniversary. There are only so many things you can do in a tent for sixteen hours, even with twenty-five years of practice.

In that same film *Out of Africa*, still on their Hollywood camping trip, Robert Redford gives Meryl Streep one of the more erotic, if fully clothed, shampoos in the history of motion pictures. The long blond tresses swirl easily into towers of suds; cold soapy water does not seem to trickle down Streep's neck; pine needles, deerflies, and other assorted bush trash do not get into her still-wet hair. Redford is somehow able even to manage to rinse it clean, judging by its later glow in campfire light. It is an act of easy intimacy, spawned, one supposes, by more of those practical realities of wilderness living: she did not, after all, bring her hairdresser on safari.

Never having given or received an erotic shampoo under wilderness circumstances, I can't vouch for the practicability of the scene. (Besides, it's no longer ecologically correct to use soap in the wild.) I do know that softness and luxury are unlikely concomitants of extended trips outdoors—except perhaps, again, by contrast. If you've been cold and wet long enough, the warmth of a dry sleeping bag is certainly a luxurious change, if hardly the stuff of erotic fantasy. Attempts at the erotic tend to come a-cropper anyway, over grit, insects, roots, rocks, and other realities. In fact as every wilderness traveler knows, the first sybaritic thrill upon returning to civilization, even before the hot shower, is sitting down again in a real chair, with something to lean back against.

This usually takes place when you regain your car at the takeout. Then most people realize they haven't seen their own face in days, and take a look in the rearview mirror. Redford or Streep ain't in it.

Luxury camping is going to a nice place and having what you need to enjoy it. The luxuries are the air, the sounds, the smells, the visuals. The luxury is the place—and, for want of a better term, the stillness, which is the best thing about the woods. Even when things are entirely noisy, when all hell is breaking loose, there is an underlying quiet. It is the most restful thing I know. You can never manage to stay continuously attuned to it, but it is always there, and you find yourself checking back into it, reconnecting with it, oftener than you expect.

At times you become so immersed in it that any movement, anything unusual, any break in the stillness itself, *snaps* your attention to it, pulls you absolutely out of yourself. One gets a glimpse then of the level of attention that wild things must bring to their surroundings: no woolgathering, no musings. No interior life at all, no time for it; all of the available consciousness otherwise engaged. Perhaps in the so-called lower species evolution has selected away from the interior life, selected for attentiveness. Where did it go so wrong with us?

The stillest times are at dawn and sunset, when it's easy to imagine that something like the spirit of the place is trying to emerge. Sensory values flip on you, the stillness no longer an absence of sound and motion, not an absence of anything but a phenomenon itself. To stop in the woods to listen and hear "nothing" is to miss the point. What you hear when you stop in the woods is everything, even if it comes in perfect silence. As with all good religions, there's nothing you can quite put your senses on, but there is a kind of signal, an idea. Not quite a presence. Perhaps the idea *is* the presence, in which case the impulse to genuflect upon stepping out the door—which has occurred to me on certain meteorologically or otherwise extraordinary days, white-pebble days—is not unfounded.

It was the stillness that made the riverbank so nice, while we were studying the moon and what were not, after all, British soldiers. It was very comforting. One needs relief from an interior life now and then, I think. Or maybe it wasn't stillness at all. Imagine a thousand monks in a vaulted stone temple, intoning simultaneously the lowest note they can sustain, a great basso profundo hum. As things turned out on that trip with the Todds, I did not think consciously of Jud at all. Dealing with the physical world—as it comes to you—blessedly militates against that part of the consciousness where concern for past and future takes place. But the past was always there, in that hum, supporting and coloring every moment of the weekend. We had promised our friends a weekend of birdsong; what we got instead was more in the nature of the forest's own Gregorian chant.

9

how a canoe works

We take our coffee down to the shore to sit on the trunk of a fallen tree where a small creek tumbles down the hillside and spills into the lake. Morning mist is spewing from the end of the stream, pouring out onto the lake surface as if from a fog machine. This puzzles me. At first I think perhaps the current from the creek is somehow driving the mist, but I can't imagine how flowing water could have that much effect on the air above it. Then I realize that the creek occupies the cleft between two steep hillsides. Cold air comes pouring down those hillsides by sheer density, funnels into the creek channel, and spills on downstream. The chill air condenses the moisture that rises from the creek into mist. It *is* a fog machine.

Before Chris and I went on our long Adirondack expedition, we were invited to go on a week-long Canadian trip with two other couples. The trip was organized primarily by Jo Currie, friend of Chris's brother Mike McCall and an experienced wilderness canoeist. Jo and Mike had their own canoe, but the other four of us used rented ones. One of the rentals was a wide-bodied tripper, with large carrying capacity for expedition purposes; the other, a cruiser designed more for speed than capacity, was identical except for a narrower waist. As the less-experienced canoeists, Chris and I took the wider, more stable boat.

It was considerably more work to paddle than the other canoe, although we didn't realize that until we'd done forty or fifty miles in it and then swapped boats for a day. Switching to the narrower canoe was like going from a dump truck to a racing car. At the end of the day our conscience made us return it, although our Canadian friends claimed they didn't notice the difference. (Canadians are born paddling, and will not admit difficulty or discomfort if they're drowning.) The next day's paddle, in what we'd come to regard as a sluggish scow, was exhausting and interminable. It insured that when we bought our own we would choose a slim cruiser.

That's the way it is with equipment: frustrating. In any effortful pursuit someone is always coming up with another piece of gear, a slightly better technology, that offers to reduce the effort—or increase the pleasure, same thing. Often it does. Good equipment spoils you very quickly. After you get used to it, bad equipment seems to add so much to the workload, which is always heavy enough. Purists would have you compensate by means of skill, which is slow to acquire, which you never quite finish acquiring, which can be more frustrating than the equipment itself—and which you can't just go buy. Most people who buy good equipment are trying to buy skill. I certainly am.

Paradoxically, the workload is a large part of why one is out there participating in the activity in the first place, whatever activity it is. This is an uncomfortable little secret I have trouble admitting even to myself. As rational creatures we are supposed to be committed to reducing the effort. The whole idea is never to have to strain at things, isn't it? Yet the truth is that we like effort as much as we hate it, perhaps because something in us knows it keeps us alive. This is entirely too Nietzschean a view for my tastes, but it happens to be the politics of the physiology.

The *Tao Te Ching* teases with the Least Effortful Way in the same way that Zen teases with Just Hauling Water. These philosophies, or perhaps they are psychologies, need concern themselves with such matters only because while our intellect wants to do things the easiest way, something else in us requires that we put more effort into them. This isn't type-A behavior, or adult-children-of-alcoholics behavior, it is only humankind at work. We're either trying hard at things or squirming in frustration because for some reason we aren't able to try hard at them.

The canoe is a long, slim boat with a rounded bottom, which to the noncanoeing world means only that it tips over easily. Capsizing a canoe is great fun when you are young and hormonal but frightening when you are not and are beginning to perceive the world as a more dangerous place than you used to. Both responses are correct. To canoeists (and I wish there were another generic term that didn't sound so much like 1880 and bloomers on bicycles), tippiness is a nonissue, although safety and self-rescue and what might be called expeditionary conservatism definitely are. Canoes aren't tippy if you are comfortable in them. It doesn't take long to become comfortable in them.

All else being equal, the longer a boat is, the faster it goes, for physical reasons that need not detain us here because I don't entirely understand them.* That's why canoes are long. They're slim because all else needs to be equal; turn that upside down and it means the slimmer boat is faster because it is longer for a given width and a given application of power. The rounded bottom is the fastest shape because it has less drag. Therefore canoes are long, slim boats that have rounded bottoms in order to go faster, which is crazy because water is sluggish stuff when you try to hurry through it but an easy, almost frictionless medium if you don't. Canoes and bicycles are remarkably efficient modes of transport, but so is walking. Humans are obviously slow-moving creatures, built to plod, pick their way, ruminate. We're closer to cows, metabolically, than to deer, yet we are determined to race each other. See condition, comma, human.

Ah, but the faster a boat is—all else again being equal—the less power it will require to attain a given speed, which is where the canoe's great efficiency comes in. What you want is *glide*—distance covered per paddling stroke. Glide is what makes paddling a canoe such serene, easy, languorous work.

*According to one guide, the maximum efficient hull speed is 1.4 times the square root of the waterline length. Let's see, length = 18 feet, square root = 4.24, times 1.4 = 5.94. But 5.94 what? Feet? What's that got to do with speed? The same figures for a sixteen-foot hull give 5.6, so the most efficient speed for an eighteen-footer is 0.34 something or other faster than a sixteen-footer; roughly 18 percent faster, anyway. Sounds to me like a gross exaggeration.

Glide is why you can go so much farther in a day than you can walk, and take along five times more weight than you could possibly carry. It is not why you are so tired at the end of that day, but it is why you are so much less tired than you would have been if you'd walked and carried the load. You're tired because you went so far and, being foolish, or human, kept at it too hard.

Of course between bodies of water you do walk and carry the load, and carry the boat in the bargain, which is why non-canoeists, after they've overcome their concern about tippiness, begin to worry about portaging the damned things. They may even have tried to lift one once. It will have been either a water-soaked wood-and-canvas fossil or an aluminum tub, in either case eighty or ninety pounds of elaborate unwieldiness. Canoes no longer need to be heavy, but light weight is expensive. In equipment, lighter weight always is. You do not buy canoes by the pound.

Actually, not all canoes come with rounded bottoms, and people do not stick with eighty-pound canoes entirely out of cheapness or nostalgia. Canoes made out of polyethylene, or a plastic called Royalex—sometimes known as "rubber canoes"—weigh a great deal but will take an enormous amount of abuse. They're not much fun to paddle and less fun to carry, but for severe conditions—facing a great deal of whitewater in places where a busted canoe means a five-hundred-mile walk, as in the Far North—they're the vehicle of choice. Canoes intended for fishing often have flat bottoms so you can stand while casting, which is frowned upon if not impossible in less specialized canoes. Unfortunately, in bad conditions a flat bottom is less stable than a round one; tilt a canoe not too far off its flat bottom and it suddenly flips. A round-bottomed canoe feels squirrely when you first get into it, but it is as stable tilted as it is upright, giving a lot more room for correction before disaster sets in.

Canoe bottoms are never perfectly flat or perfectly round anyway, but a compromise between the two, the curve of the round bottom slightly flattened out, or given a shallow vee, then rising into more or less vertical sides. There are other sophistications. Flared sides—curved outward at the gunwale—make for a drier boat and add resistance to capsizing; sides that tumblehome—curve back in toward the center of the boat at the gunwales—allow the paddler not to have to reach so far to paddle, which is not only more comfortable but puts the application of force closer to the line of travel. The closer the paddle stroke is to the center of the boat, the less steering correction will be needed.

In fact, as with any other human contrivance intended for use in the physical world, everything in canoe design is a compromise. Long boats go straight better but are therefore harder to turn than short boats. More freeboard makes a boat drier and gives it more carrying capacity but makes it a handful in windy conditions. A wider boat is more stable and carries bigger loads but, since it pushes aside more water, requires more effort to paddle.

One of the reasons that long boats are faster than short ones is that the hull of a short boat widens more abruptly, pushing water aside at a sharper angle, which creates more drag. Furthermore the shape of the hull—how sharp the bow and stern are, how quickly the hull widens out, where the widest point is along the length of the hull, and other subtleties—involves some rather dark science. The matter of what volume you put into the water and where you put it does not seem to be reducible to simple laws, which is why naval architects get to spend so much time playing with little toy boats. More marine design than one might expect, in a computer age, follows the scientific principle of TLAR, which is technologese for That Looks About Right.

Horses for courses, in other words: you choose a canoe for the use to which it will be put, because no canoe does everything well. Recreational canoes are stable platforms for leisurely use on fairly calm water, and, I suppose, fishing, although I am not a fisherman. Touring canoes are for carrying a reasonable amount of gear for reasonable distances through most conditions. Cruising canoes are for going fast or far for the pure pleasure of paddling. Expedition canoes are for carrying big loads through all possible conditions. Racers are for racing.*

Different canoes for different uses brings us to the matter of whitewater. The straighter the line of the keel in relation to the waterline, the better a boat tracks, which is precisely what you want for covering distance, but the more reluctantly it turns, which is the opposite of what you want for dodging rocks in rushing water. Whitewater canoes are shaped like bananas, bent in the middle for maneuverability. They turn instantly, at a touch of the paddle—so much so that it takes an expert paddler to make one go in a straight

*This rather simplistic discussion of canoe design is based heavily on a much more detailed essay on the subject in the *We-no-nah Canoe*, volume 10, the catalog published by We-no-nah Canoe, Inc., of Winona, Minnesota.

line at all. They are for playing in rough, tumbling water (or for slalom racing).

Whitewater canoeing is a sport for hormonals. Having been hormonal myself, I know the attractions of such thrilling activities, but now sometimes find myself wishing there were a clearer separation between them and me. The hormonals seem to need to use the physical world *hard*, leaving their mark on it, even if only in the form of noise; the less hormonal among us, for whom being there is enough, are almost apologetic for our intrusion. Hormonals don't much mind nonhormonals being around so long as the nonhormonals don't get in the way. Unfortunately, it doesn't work in the other direction. Now that I am no longer so hormonal, I complain about those who are. The *Tao* tells us we have to accept it all, but how could I have been like that?

In the matter of equipment, you can allegedly overcome deficiencies of design with good technique. The folklore of sports is full of stories about the expert who makes fools of the rest of us, using bad equipment. There used to be a golf hustler who could drive the ball with a soda pop bottle and still take your money. At the Sebring twelve-hour motor race I once watched Stirling Moss put what was essentially a street-model sports car up among the overwhelmingly faster pure racers for several hours, until the poor overworked sports car finally fell apart. Good technique can overcome a lot.

Good technique is never anything more than doing things the easiest way possible. Therein lies a strange difference between practice and performance, or between training and competition. In training you do things the hardest way possible, in order to load the systems and force them to adapt if they are living systems, or to reveal their weaknesses if they are mechanical ones. In training you spend as much energy, or apply as much force, as you can. In competition—in any kind of actual performance—you do things the easiest way possible within the framework of what you are trying to accomplish. You save as much energy as you can while fulfilling the requirements of the event.

A lot of us find that doing hard things the easiest way possible is, after sex, the most fun one can have with the physical self. It is an engagingly subtle task. It requires learning how to rest, how to take little vacations during the expenditure of effort, how to see through to the end of the job: how to get that shovel of dirt out of the ditch with the least possible expenditure of effort, even if it's only in order to shovel more dirt.

Circumstances once required me to move a stack of firewood into my basement at a time when I had a bad shoulder. It was an interesting problem: to lift, haul, and stack the weight with a minimum disturbance of the flesh. By the time I finished I was paying attention to every little detail of the task, even to breathing and balance, trying to *ease* those chunks of wood down the steps and onto the stack. Pain, like fatigue, teaches efficiency.

I caught myself thinking that it was a lazy way to work—a curious hangover from the teachings of my boyhood. Where I grew up, "lazy" was white middle-class shorthand for what might be called an ethnic style of manual labor. One heard a lot of racist folklore: just watch them work, look how slow they move. "They" cultivated a plodding pace, minimal movement, dependence on gravity and momentum and leverage to do the work instead of more ambitious muscle.

Their shovels never cleared the ground by an extra quarter of an inch.

I secretly wanted to work that way myself, as any fool might, but it didn't look right for a white boy to emulate "them." I'd get chewed out for it. I was expected to work at a white boss's frenetic pace—efficiency be damned, let's set an example—and the habit unfortunately took hold. I never managed to acquire the temperament.

I relearn this every spring when I begin trying to get our place back in shape. A few years ago I set out to clear a small patch of grapevine-entangled trees between two fields, a job I optimistically estimated would take half a day. In four hours I had cut perhaps a tenth of what I had in mind and was already working in a kind of growing hysteria, fatigued into stupidity.

I'd sawed through a group of five-inch trees that snagged at a forty-five-degree angle, their crowns so entangled they couldn't fall. I figured I would monkey up one of their slanted trunks, bounce up and down on it, and bring them all sliding gracefully to earth. When I finally fell out of the tree, the wrist of one of my work gloves caught in a forked branch, leaving me unhurt but hanging by one arm, feet a few inches above the ground, sweat pouring into my eyes, blackflies beginning to feast on my exposed rib cage as I tried to figure out how to get myself down. I reflected then on learning to work in a somewhat lazier fashion, and I'm telling this now for what the psychologists call reinforcement. Chris suggests I tape a copy to the chain saw.

Like the swimmer's hand (and foot) and the canoeist's paddle, the bird's wing is also a device for the application of force, not to mention exploration of fluid dynamics. It is equipment. Walking in the woods at dusk, I realize things have gotten very quiet, and stop to listen. A large owl lands in a maple thirty feet above my head. It watches me for a while, checks out the scenery, then flies twenty yards to another tree, lands, surveys the scene again. It turns its back to me. Unable to contain myself, I make clicking noises, calling to it. It ignores me at first, then turns to look at me again. Then it flies, calmly, to another tree, again about twenty yards away.

I had heard about the silence of owl flight but never experienced it before. From a bird that large, one expects wing noise, air noise. The specialized plumage of owl wings just seems to catch the air, hold the air; the bird flies, effortlessly, almost as if parachuting from tree to tree. It is wafted; it waves these soft, fluffy things and is lifted up, as if on a string, and flown to the next tree. The action is curiously different from ordinary bird flight, so different from upland game birds, which explode out of cover and make you jump out of your skin.

In the owl's case, the experts tell me, the difference is equipment rather than technique.

10

rock and water and sky

> Water can be both good and bad, useful and dangerous. To the danger, however, a remedy has been found: learning to swim.
>
> —Democritus

Lying back against a voluptuous swirl of smooth pink rock, I idly watch a skyful of soaring birds, trying to decide if I've baked long enough in the late July sun to make the ice-green water at my feet once again gratifying, rather than mortifying, to the flesh. The granite finger onto which we've pulled our canoe is separated from the next island, perhaps even more inviting than the one we're on, by fifty yards. We'll check it out in a minute, or maybe tomorrow. We have roughly 29,999 more islands to choose from, and seven days in which to explore them—if we decide to, instead of just lying here and swimming and watching the birds. I find myself thinking that we've finally succeeded in getting away from the world, and then realize I've got it backward. The world—the phenomenal world, unmediated—is just what we've found our way into.

We are on the southern edge of the Precambrian Shield, in the maze of granite where the French River loses its riverine identity and gradually turns into Georgian Bay, which in turn becomes Lake Huron. We are also at a major checkpoint on the oldest voyageur route of all, for over two hundred and fifty years the most heavily traveled canoe route in North America. Nowadays it is merely one of the most beautiful, and we have it virtually to ourselves.

Geography *is* destiny, never more persuasively so than in Canada. That nation's central core, its defining fundament, is the Shield, the overturned frying pan of ancient granite that stretches from the Great Lakes to Hudson Bay and from the Labrador coast to the shores of the Arctic Ocean. Because of the Shield's hostility to agriculture, Canada didn't draw settlers, it drew traders bent on extracting its resources; because of the Shield's hostility to travel, those traders historically kept to its rim, leaving its interior, except for a few rare geological corridors of access, to the native population. One such corridor—the most important—was the French River.

For canoeists, fortunately, the Shield is not so much a barrier as a great cistern, storing in pristine lakes much of the water it collects, spilling the rest down the rocky channels that form its thousands of heart-stoppingly beautiful rivers. One half of the freshwater on the surface of the earth is in Canada. Because of this the canoe became roughly as important to its history as the horse was to the American West.

A few miles north of where the Shield begins to break up and disappear into Lake Huron, it is transected by a sharp east-west fault line that now forms the bed of the French River. Down that river in 1610 came the first European to penetrate the interior of Canada, a sixteen-year-old named Étienne Brûlé. He had been sent by Samuel de Champlain, who'd inflamed the Iroquois in the Adirondacks the year before but who was now busily trying to shore up the tottering French colony at Quebec. Brûlé's assignment was to go to live with the "Savages" and learn their languages. The French River was hardly a new route, having already seen several thousand years of canoe traffic, but in the seventeenth century it was the only passage between the Saint Lawrence River and the Great Lakes that avoided those French-hating Iroquois: up the Saint Lawrence to Lachine Rapids, at what is now Montreal, where the Ottawa River comes in; up the Ottawa to the Mattawa; up the Mattawa to Lake Nipissing; across Nipissing to the French River; down the French to Georgian Bay, the Great Lakes, the entire western fur country.

You can still canoe that route—but then you can canoe from almost any Canadian city to any of that country's ocean boundaries, including the Arctic.* Most of the old route, however, has been overwhelmed by settlement. The Ottawa River, heavy with traffic and dammed into a series of reservoirs, is no longer an appealing paddle. The Mattawa is attractive enough but short, crowded, and seldom out of earshot of a busy highway. Lake Nipissing is shallow, choppy, blighted on the east with heavy industry and a city, North

*This is literally true—and the longest portages required are twelve miles at Methye into the Mackenzie Basin and nine miles at Grand Portage out of Lake Superior's western end. Few other portages are longer than a couple of miles.

Bay. The French River, though, is still much the way the voyageurs left it. It is the sort of waterway for which the North American canoe was, literally, invented.

Our group met on Georgian Bay, making a late-afternoon speedboat run through an intricacy of low granite lumps out to the island cottage, belonging to a painter named Geoff Armstrong, where we would spend our first night. Geoff and another painter, Chris Adeney, knew the French River well and were our putative guides, but the trip had been primarily organized by Jo and Mike, who had done it the year before and were eager to go back. There would be three canoes: two couples and two guys batching for a week; a sojourn, not an adventure. I was the sole non-Canadian.

On the wall of Geoff's cottage hung a large aerial photo of the area: several thousand white dots on a black sea, the fractionated lower edge of the Shield laid bare—and a perfect forewarning of a terrain that cannot be grasped without detailed maps. Fault lines cross everywhere, giving a mishmash of straight-line rivers dissected by diagonal slash marks, so knitted together as to give new meaning to the term reticular. Glaciation and erosion have combed much of the granite into long, thin, north-south fingers that make narrow interconnected inlets and passageways, delectably navigable by canoe. The shoreline is sometimes sparsely forested, but exposed rock dominates the landscape. It is rock one cannot take one's eyes off. It is what we gazed at, after dinner, as late July sun set in a clear sky behind low scrub pines. Painterly discussion arose about the true color of the backlit shadows, but nothing was resolved. By dark all was water and sky and rock.

• • •

Five years after Brûlé, Champlain himself covered the same route, as did four Récollet fathers bent on converting the Hurons of Georgian Bay. Other Récollets followed, six of whom would drown at what is now called Récollet Falls. Jesuits—the Black Robes of Brian Moore's novel and the film of that name—followed the Récollets, spreading the True Cross, until smallpox, assisted by Iroquois, effectively wiped out the Hurons.

Unfortunately for the native groups that survived—not to mention the fur-bearing mammals of North America—Europeans now knew the route. Nicolet used it on his way to "discover" (lay white man's eyes on) Lake Michigan in 1634; Marquette and Joliet, La Salle, Radisson and des Groseilliers, La Vérendrye all used it for their explorations. The *coureurs de bois*—freelance French fur gatherers, forerunners of the voyageurs—used it heavily.

Eventually the route became what historian Eric W. Morse called the "Montrealers' Mainline": the conduit through which the voyageurs tied the fur country of the Athabasca to the markets of Montreal. "This 'Trans-Canada Highway No. 1,'" Morse writes in *Fur Trade Routes of Canada*, "was Canada's economic spinal column—and even after 1821, when the western freight traffic shifted [to Hudson Bay], remained the only route for mail and passengers to cross Canada until the railway came in the 1880's."

The French River is only seventy miles long, which was an easy day's run for the voyageurs, even with a couple of portages. A good part of the river follows that east-west fault line, making a startlingly straight run of several miles between stately granite cliffs. Our painters no longer found that long straightaway entertaining—particularly when prevailing winds can turn a fifteen-mile haul into a day-long struggle—so we put in at Hartley Bay, a little above the point where the river turns south and heads for Georgian Bay. After paddling on slow, dark water for a leisurely couple of hours, we camped at an unfortunate intersection for outboard traffic called Crombie Bay. Except at the put-in, we had seen no other canoes.

The French was the first Canadian river to be given Heritage status, analogous to the Wild and Scenic system in the United States. Much of it is de facto wilderness, essentially unchanged since Champlain saw it in 1615, despite the fact that its eastern end cuts through an Indian preserve, the surrounding pine forests were logged heavily around the turn of the century, and cottages dot the river's banks here and there. Its stony passages contain prehistoric pictographs; until not long ago, bison and elk grazed its northern edges. In its lower stages it is a maze with four mouths, not to mention those hundreds of crisscrossing channels and thousands of small rock islands extending out into Georgian Bay. The bay was where westbound voyageurs, and we, were headed.

Of the four mouths of the French River, the voyageurs preferred the Western Channel, seeking shelter as long as possible from the open waters of Georgian Bay. In particular they dreaded rounding Point Grondine (French for "growling," for the noise when winds batter the point), exposing them to the long fetch off Lake Huron. They would often stop for a few days in a certain distinct meadow a couple of miles short of Grondine to wait out the weather.

Current maps, confusingly, show both a Voyageur Channel and an Old Voyageur Channel. The latter, an extremely narrow and twisty pathway with one portage—a lift-over, really, of no more than ten yards—was the historical route of choice. Voyageurs called the portage "Petite Faucille" (little sickle); in high water it made a navigable curved rapids from which modern divers have recovered fur trade–era hardware. The voyageurs had expanded the native birchbark canoe—a more efficient means of wilderness transportation than any European would ever devise—to its physical limits, in the form of the gigantic thirty-eight-foot *canot de maître*, paddled by a crew of ten. It is hard to imagine a thirty-eight-footer passing between the high rock walls above and below the Petite Faucille, but contemporary accounts insist that the big canoes breezed through. Finding the old channel is not easy—topo maps and compass are a necessity—but worth the trouble to tease it out. Rock walls shut out wind and noise, the tortuous route slows down the paddle strokes, and before long the imagination drifts back through the centuries, waiting for a boatload of voyageurs to round the next bend and call out for right of way.

It must have been an inspiring sight. A nineteenth-century account by R. M. Ballantyne describes it: "I have seen four canoes sweep round a promontory suddenly and burst upon my view, while at the same moment the wild romantic song of the voyageurs, as they plied their brisk paddles, struck upon my ear; and I have felt the thrilling enthusiasm on witnessing thirty or forty of these picturesque canoes . . . half shrouded in the spray that flew from the bright vermilion paddles."

The rock is dazzling. Sculpted smooth by millions of years of advancing and receding ice, it comes in voluptuous pink sworls interspersed with gray or charcoal black: sexy, soft looking, something like Navajo sand paintings surrounded by water. Dead-level yellow pollen lines, sometimes stacked six or seven deep, mark previous springtime water levels. Where the smooth surface is cracked a darker red is exposed, contrasting to the sun-baked pink. Here and there a harshly rectangular chunk has fallen out of the smoothness, putting abrupt, hard-edged shadows in the midst of the polished complex curves, as if a brick had fallen out of the fender of a Porsche.

Add to that jade green water, growing steadily clearer and colder as we drew nearer to the bay. Okay, I'm a sucker for clear water over smooth rock: the old channel is the best place I've ever been in a canoe. We gawked our way through it, then turned due west and picked out in the distance an unmistakable softening of terrain that marked the voyageurs' meadow campsite.

Amateur historians originally identified the meadow by a couple of long, low, parallel mounds, apparently manmade—workbenches, speculation had it, for canoe repair. The mounds are still there, impressing us all, although Chris, as expedition logician, wondered what was gained, accesswise, by laying a canoe that was already five feet tall across eighteen-inch lumps of earth. The voyageurs' meadow has obviously become a modern duck hunters' rendezvous, littered with spent shotgun shells, tattered plastic, beer cans, and assorted other detritus. Clots of river grass and pickerelweed guarded its muddy shore. On a hot afternoon it was a moderately pestilential place, swarms of deerflies disrupting our flights of historical imagination, and we were glad to get back on the water. The open waters of the bay were only a couple of miles away.

• • •

We'd swum in dark, cool waters at sunset in Crombie Bay, and before entering the old channel we'd stopped for lunch and a reinvigorating swim. After leaving the meadow we found a campsite of our own and were quickly back in the water—now not to be resisted, clearing to an almost glacial Huron green, chilly enough that to immerse was a thrill but to soak was bordering on the hypothermic. It made for highly energetic, spirit-raising swimming.

Finding a campsite had meant wandering around confusedly among indistinguishable whale-back islands that looked for all the world like mock-up pink submarines designed by Aubrey Beardsley. We were searching for a site that Chris Adeney remembered fondly but couldn't locate. An osprey flew over with a fish firmly in its grasp, circling, circling, as if it couldn't find its nest either. Although Cdr. Mike McCall (ret.) had brought nautical strip-charts even more detailed than our topos, and could always tell us where we were, the problem was where we ought to be going. We didn't find the old site but eventually discovered a better one, our own personal pink submarine, and settled in. A sprinkle of rain stimulated Armstrong, an architect as well as a painter, to erect one of the more complexly elegant tarpaulins in the history of camping. Adeney produced not only a shrimp paella but, to general applause, a nalgene bottle of white wine. Nobody had said anything about traveling light.

Sunset was nondramatic for a change, a lemon band on the horizon under blue-gray cirrus. In that setting if you weren't looking at water or rock you were looking at sky, which almost always featured at least one fairly large bird: osprey, gull, cormorant, tern, hawk. Lots of turkey vultures.

If large, the bird would be doing some fairly fancy flying in its pursuit of whatever its avian desires happened to be. They were wonderful to watch.

Georgian Bay sent a rattling thunderstorm overnight—always a thrill in a tent—and a foggy morning, leaving us damp enough to vote for a layover. Besides, the place where we were camped was too pretty to leave. While the painters painted, the rest of us paddled back up to the main channel, just touring, moderately lost for a while until we met up with four women who'd canoed in from Killarney Park, twenty miles west, and knew exactly where *they* were. As we headed back for lunch the sun began to burn through, generating a lifting fog and a strange and lovely light that had the painters marveling, reminding us to marvel too. The swimming improved.

In our absence the painters had also gone exploring, and had found a sheltered passage out to the surf line. After supper they took us back, leading the way through a labyrinth of hidden channels eight and ten feet wide, down long corridors of sensuously swirling rock, at least one of which was marked with a spray of wild roses to help us find our way back. We ended up at a granite headland overlooking pounding surf, our first real contact with the immensity of Georgian Bay. Not for open canoes, at least not that evening. There'd been talk of making our way out to the Bustards, a group of small islands a couple of miles offshore, to spend a night if the wind dropped. Otherwise we'd leave that to kayakers.

On another long twilight Adeney would lead us to an uncharacteristic broad stone plateau, close by the bay, in search of ancient camps. Tradition has it that local tribes gathered there in springtime to dry their fish, choosing these windy flats for relief from insects. We found several of what we assumed were tepee rings, in place long enough, as Adeney pointed out, for lichens to outline the stones that outlined the circles. There was no trash, which is perhaps a better indication that the stones hadn't been put there by moderns.

It is entirely satisfactory, and haunting, to stand on wind-blasted rock, casting one-hundred-foot shadows in the summer dusk, in a place where people have been camping out for thousands of years. (A startling thing about wilderness is that it is sometimes not about wildness but about history: look, this—explicitly—is how it was, because it still is.) If not Hurons they were probably Ojibwas, who before European contact were using copper from the north shore of Lake Huron and are assumed to be the ones who left the pictographs along the river. The Ojibwas, of Algonquian stock, eventually became middlemen, trading with Europeans at the Saint Lawrence end and with Hurons at the Great Lakes end. The Iroquois controlled the upper Saint Lawrence and the lower Great Lakes, and in 1649 began the campaign that scattered the surviving Hurons. They fled in all directions, some ending up, by 1867, as far away as Oklahoma. About five hundred contemporary Hurons still live on a reserve near Winnipeg.

It was a bloody business. The Hurons hated and feared the Iroquois. When a Huron sneezed, his buddies didn't say "Gesundheit" but something that roughly translated as "Fuck the Iroquois." Although peaceable, the Hurons were a fairly vindictive people. One missionary reported that they ate the lice they plucked from their hair, not for the nutrition or because they liked the taste but to get even with the lice for biting them in the first place.

I couldn't help imagining Hurons, sitting on this or that beautiful rocky point, watching the voyageurs storm up and down their river. They must have felt the same way we did on Crombie Bay, watching outboards whine back and forth through the sunset.

All of Georgian Bay is not the canoeist's private preserve. In our wanderings we picked our way past Bad River Bay, a popular gathering point between the park and the bay proper. Getting there meant crossing the foot of a series of rapids that tempted us to play, even fully loaded—but motorboats run these chutes regularly, and, once committed, do not give way to flimsy canoes. One recreational rapids runner, roaring demonically down one chute and back up the next, was a teenage boy in a tiny homemade scow with far too large a motor on the back, accompanied by a considerably younger female child and an unwilling white house cat, yowling and straining at its leash. "An odd planet," says Polish Nobelist Wislawa Szymborska, "and those on it are odd too."

Bad River Bay held several million dollars' worth of yachts, tucked away against foul weather: the Mackinac Straits race had just finished, and more thunderstorms threatened. Big boats sat at anchor running their generators to power their air conditioners. You couldn't hear yourself think; we averted our eyes and kept paddling. Finally another lift-over, a granite shoal that incontestably barred powerboats, returned us to tranquility. Inky backwater gave way again to the chilly green of the bay. The sun came out, and we spotted such an attractive campsite, on a low stone spit within sight of the breaking surf of the bay, that we quit for the day. It was barely past noon and we'd come only seven miles from our previous camp, which was a little embarrassing, but then we hadn't come here to break any speed records, had we? We set up tents, sunned, snoozed—drunk on physical geography. The swimming improved.

That was roughly how the days would go: a few hours of paddling, followed by a leisurely negotiation about campsites, there being sufficient choice that connoisseurship became an issue. (For our tentside swimming, do we want to dive from one-foot ledges, three-foot platforms, ten-foot cliffs?) The painters painted, the nonpainters wandered, by canoe or on foot. At some point in the day we might paddle out into the swell for a sniff, dodging back inland when it got too rough. Napping was institutionalized; evening meals were large and ceremonial, usually followed by an after-dinner expedition. The country began to flatten, the sparse shoreline vegetation knocked low by exposure, the fingers of island separated by wider channels, the ubiquitous rock now sometimes black or green instead of pink. At one time I found myself thinking—lolling again on a private stretch of satiny rock, nude—that my life had worked out far better than I'd ever dreamed it might, simply on the evidence of being here.

One last piece of history beckoned on our way out. Récollet Falls turned out to be more pretty than ominous, dropping eight or ten feet in three main channels; kayaks might conceivably run the falls but open canoes probably should not. Certainly thirty-eight-foot birchbarks carrying three-ton payloads should not. Native canoes carrying terrified priests dressed in heavy robes and wooden sandals definitely should not. No commemorative plaque or other memorial marked the spot, at least not that I could find. The portage around the falls is now a fifty-yard boardwalk, wide enough to allow small powerboats to be winched up and down—but by then the river had become a little too civilized for comfort anyway. The last couple of miles were outboard infested, the takeout a mixing bowl of pleasure craft and fishermen. After a week of truly fresh air, it's amazing how powerfully the exhaust from internal combustion engines stinks.

Two things about the French River remain strongest in memory. The first was that we expected to have our heads full of the romance of the voyageurs, but what I kept thinking about were the Hurons. It was their country, not the voyageurs', and I found myself seeing it through Huron eyes. The "history" of this extraordinarily historical area, even as whites recount it, is a series of intrusions. Once you get attuned to the character of the place, that's the way it feels. The historical connections are interesting, but you can't help wishing they had never happened. Except then of course you might never have gotten there.

The other memory was, simply, rock. From time to time the painters' canoe would get ahead or fall behind, and when we rejoined them we'd find them staring at some stupendous formation or other, making vague hand gestures as they struggled to take in, to preserve in memory, the immensities they were seeing. All of us talked about the rock all the time; nobody managed to say anything particularly coherent about it. It was, as someone once characterized the snow avalanche, a powerful stronghold against analysis: rock as consonant to the water's vowels.

The Tai Chi master suggests you hold *ki* energy like an imaginary ball between your hands, playing it back and forth, to get to know it. I'll never experience that, but this was the first trip on which we paddled long enough that I began to feel the enticing sphere of energy in the water beside the boat—all the water I could comfortably reach with the paddle—and got the sense that by shaping it, pushing and pulling against it, carving it, molding it, I could make the canoe go where we wanted it to go. Playing with that added fundamentally to the sensual pleasure of the trip.

French River was nothing if not sensual, bombarding us with color and shape and texture: phenomena brought to us courtesy of light. One mustn't forget the light. The reason to look at rock and water and sky is for what the light does. Out there on those exposed stone points, light gets *at* you, brings itself to your attention, in whatever forms or values it has chosen for the particular moment—and it was always changing, is always changing, which is why it demands attention. The goal for your retinas, after all, is light. Why else would the painters, whose interest is not in things so much as how light reveals those things, be so fond of the French River?

I discover this only afterward, reconstructing the trip in my mind, and wish I'd thought about it more at the time. A principle suggests itself: no matter how significant an experience of the physical world you're having, it will be intensified, and improved, if you pay attention to the light. It is the best way I know of staying in the present tense.

11

the absence of water

> Life and water are inseparable: 70 to 95 percent of all fresh fruits and vegetables are composed of water, 70 percent of all human beings, and the chemical reactions that sustain the lives of every organism take place in aqueous solution, most involving water as a reactant—photosynthesis constituting but a single example. Eighty percent of the earth's surface is covered with water in solid, liquid, and gaseous forms, its total amount being estimated as 1.33×10^{24}kg—or 5 percent of the planet's total mass.
>
> —DAVID JAMES DUNCAN, *THE RIVER WHY*

I suppose I've been chasing after perfect waters all my life, pursuing some platonic ideal, some holy grail of liquid perfection, although I doubt I'd recognize it if I found it, not even being sure what its characteristics, other than clarity, might be. I lived for a while in desert country, and sometimes amused myself while waiting for sleep by inventing more watery landscapes. Places to swim. I was not terribly original. I never got much beyond some kind of waterfall into a deep pool, with cliffs for diving and perhaps fruit trees for shade, dropping plums and oranges into my lap. Sometimes I could almost feel warm rock under my bare feet. It was very soothing.

The desert time was in far-western Texas, a very strange place. My first wife, Nancy, and I came out of college thoroughly broke and immediately began looking for schoolteaching jobs, the only work for which we were qualified. In 1955 the beginning annual salary for a Texas schoolteacher was $2,400, so when a superintendent on a recruiting trip told us about his school out west, where taxes from local oil revenues made possible a beginning salary of $3,600, we leaped at the chance. Wait, he said, we should understand that it was pretty harsh country. We'd better come take a look before making a commitment.

We did, and were more or less aghast. Partly it was the sheer distance: two hundred miles west to Abilene, another two hundred beyond that to Odessa, and then ninety miles south to the tiny town of Iraan. The name is an elision of Ira and Ann Yates, major donors to the University of Texas's endowment funds: the late Ira G., a grocer, had discovered a pool of oil so close to the surface of the earth that there were open-pit wells still in operation when we moved there. The stench of raw petroleum filled the air miles before the town came in sight. You got used to it, we were told. We took the jobs.

Iraan sits in a northern lobe of the Chihuahuan desert, not sand dunes but rimrock country, where flat-topped mesas slope down a thousand feet or more into dried-up arroyos. The region gets eight inches of annual rainfall or less; what little vegetation there is is mesquite, catclaw acacia, salt cedar, and cactus, through which roam mostly jackrabbits, lizards, and tarantulas. The town itself—about nine hundred residents in 1955—serviced the oil patch. Just east of town is the Pecos River, which cuts here and there through red rock cliffs and is serenely blue and beautiful but irredeemably salty, its watershed the bed of the prehistoric Permian Sea. The region bakes in hundred-degree temperatures from April to October and is rocked by dust storms—sandpapering the paint off houses, stripping chrome off bumpers, etching windshields into opacity—every time a norther comes through.

And just because you're not having a dust storm doesn't mean the wind stops blowing. Spring in particular brings about six weeks of uninterrupted thirty-mile-per-hour winds, rattling windows night and day, yanking doors out of people's hands, knocking hats off. I remember once talking about this with a large, gloomy old-timer as we stood in a sheltered doorway, watching tumbleweeds blow down the road. Before the spring winds stopped, he told me with a sigh, he'd lose twenty pounds. Happened every year. Gets on your nerves.

There were openings for both Nancy and me on the small high school faculty. Geography aside, it was conventional small-town schoolteaching, at which she was quite good and I was not. The plant was modern, the facilities sound, because the school system was rich. The budget for the football team was $15,000 a year, for the band $7,500, for the library $150, a distribution of funds not untypical, at least in those days, in Texas public education. The students were children of either oil field workers or sheep ranchers. Enormous amounts of land were required to raise any kind of livestock; as I recall, fifty-thousand-acre ranches were not unheard of, so some of our students rode the school bus for considerable distances.

Having a town in such a place was patently absurd, made possible only with the development of deep-water pumps that could tap the Oglala aquifer. Fifty miles to the north of

Iraan, water wells two thousand feet deep made the famous Pecos Valley cantaloupes possible, and farther north of that, irrigated farms grew cotton, alfalfa, even wheat—and gradually drained the aquifer. Twenty years after I left I was told that to sell a farm now required certification of the remaining water supply: three more years before exhaustion, five more years. If you were lucky, another thousand feet of drill pipe might reach more water; if you weren't, the land was "useless," at least for anything the locals could think of. Current debates over nuclear dump sites come to mind.

Iraan—endless acreage with nothing on it but a scattering of smelly oil wells, a perfect wasteland—was already useless, as far as any rational, tree-oriented, water-dependent human being such as my callow self could see. Recreation and entertainment were self-administered. We were out of range of the little television that was available at the time, and radio reception—of evangelist preachers and hillbilly music—wasn't much better. We did without newspapers or any other connection, essentially, with the United States; nothing "national," other than brand names in the grocery store, entered our lives.

High school sporting events were extremely well attended. Local nimrods blasted away at the doves and waterfowl that congregated at stock tanks, at deer (a fellow faculty member was fined for headlighting them while we lived there), even at javelinas, and, when nothing else was moving on the horizon, at the resolutely inedible jackrabbits. Just for something to shoot at. (Armadillos were for some reason spared, perhaps because you could run them down on foot.) Three of the faculty were serious fishermen in an essentially waterless landscape, making an annual spring float trip down the nearby Rio Grande as soon as school was out. At the time

I thought they were crazy, but now I wish they'd taken me along, even if I didn't fish. The scenery must have been fabulous.

The faculty social life—dinners, cards, dominoes—was inbred and fairly intense. We lived in "teacherages," a row of tiny school-owned frame houses that had been moved into town from a nearby oil camp and plopped down near the schoolyard, so we were a neighborhood as well as co-workers. (Houses were moved regularly in west Texas; new lumber had to be trucked a very long way.) Rent was cheap but life was circumscribed. The school's maintenance crew picked up our garbage, for instance, and it was generally understood that if they came across a beer can they'd report it and you would be fired. Church attendance was expected, your choice of Baptist, Methodist, or Church of Christ; if you required a more exotic sect you drove to Odessa or Midland, 180 miles round-trip—but then we did that regularly for movies anyway. Being church-resistant by both nature and nurture, I attended a few Methodist services, grumpily, and then discovered I could simply drive out of town on Sunday mornings, leaving my affiliation as mysterious as possible. No one asked.

We lasted four years, fleeing in summers to moister climes, going back again after reaffirming that teaching salaries in the rest of the state had not caught up. One summer I stayed in Iraan and taught swimming lessons at the town pool for extra cash. To my amazement I was miserably cold the whole time: the artesian well water never warmed up sufficiently, even in 105-degree heat, to overcome the chill factor of desert wind on wet bodies. Instruction was a good idea, few of the locals being able to swim at all. There was not much opportunity to learn. There was a golf course next to the pool, now that I think of it, with sand "greens," à la Kuwait.

Another specific local recreation that drew its share of partisans was arrowhead hunting. There were Indian caves—really only shallow indentations under the rimrock—all over the place. You simply looked for a dark blur in the otherwise dun-colored rock, marking the spot where eons of campfire smoke had smudged the overhang. You could spot them from a mile away. When you climbed up to them, sometimes you'd find a *metate* hollowed out in the rock floor, for the grinding of grain. Whatever other signs were extant, the scree below the cave and the ledge above would be littered with flint chips. The short-grass prairie had once extended that far south, making the Pecos Valley the lower end of the buffalo range. Aboriginals sat up there to watch for bison, chipping flint to pass the time.

Or so the locals said; I'm not sure of the anthropological accuracy of this picture, but the flint was definitely there, and broken and discarded projectile points were not rare. Perfect ones were, and were highly treasured. It was a lovely way to spend idle hours if you enjoyed walking around in rough country. You did have to be careful, since the vegetation was chaparral, the mother of invention of the cowboy's chaps. One did not brush aside anything carelessly. It is not a comfortable habitat for large, clumsy, thin-skinned, essentially hairless vertebrates. (The desert motto, says Diana Kappel-Smith: "Trust nothing.")

I restricted my expeditions mostly to those fine spring and fall days when the temperature might be twenty degrees at dawn but would reach seventy by noon. More enthusiastic arrowhead hunters would go out anytime, searing heat notwithstanding. Our three faculty fishermen in particular were almost as enthusiastic about this pastime as they were about fishing. One of them, high school principal Jack Black, appeared at my classroom door one Monday morning in May and waved me out into the hall. He was practically dancing on the spot, barely able to contain his laughter. He just had to tell me, he said, about that Sunday's arrowhead expedition.

He'd gone with his two fishing buddies, basketball coach Lee Burnett and math teacher Henry McEwen, to check out a new canyon. The day was blistering hot, as usual, and they walked quite a distance. Foolishly, they hadn't brought water. After awhile they began to get extremely dry, so they climbed out on top of the rimrock to pick a straight-line route back to the car.

"We could just see it, a couple of miles away, but we also saw a windmill about half a mile off to our left," Jack said. "The blades were turning, it looked like it was pumping, so we figured we'd swing by it first, get a drink, then head on back to the car.

"Man, it was hot. We finally got all the way down to the windmill, and it was pumping away just fine, water pouring out of the pipe. Lee stepped up first, bent over and drank and drank, then stepped back and let me go next. As soon as my lips touched the water I realized it was pure brine. Lee was just putting on an act. So I did the same thing, gulp, gulp, as if I was drinking. I smacked my lips and stepped away to let Henry have a taste.

"Henry bent over and started to drink. Lee and I were standing there looking at each other, waiting for him to react. He just gulped away like we had, then straightened up, wiped his mouth, and started walking to the car. We followed him, got in the car, and drove home. Nobody said a word about it all the way back."

I laughed and went back to my class. Half an hour later Lee showed up at my door, waving me out into the hall. "I have to tell you about our arrowhead hunt yesterday," he said. I didn't get Henry's version until that afternoon. As well as I was able to determine, the three of them never did discuss the matter among themselves.

For a river that goes essentially from nowhere to nowhere, passing mostly through nothing, the Pecos has seen a great deal of history. It rises pure and fresh in the mountains of New Mexico and empties nine hundred brackish miles later into the Rio Grande at Langtry, on the border with Mexico. Billy the Kid's grave lies near one end, Judge Roy Bean's at the other. In some senses the Pecos Valley is the real geography of an otherwise mostly mythical American West. It marks the western boundary of the featureless Llano Estacado—the "Staked Plain"—which so baffled early explorers; de Vaca, Coronado, and de Soto all crossed it in the sixteenth century, La Salle in the seventeenth. Bosque Redondo, on the banks of the Pecos near Kid's grave in New Mexico, was where in 1863 the U.S. Army interned the eighty-five hundred souls that they had force-marched from Navajo homelands; ten thousand had started the trip. Five years later the reservation was abandoned and its residents were allowed to walk home.

The Pecos was also the original western boundary of Lincoln County, where Billy the Kid was blooded in New Mexico's notorious Lincoln County War. John Wesley Hardin, Jesse James, the Younger brothers, the Daltons, Sam Bass, Big Foot Wallace, and Calamity Jane were at some time in their careers associated with the region. Fort Lancaster, one of the ring of frontier outposts erected in the 1850s as defense against the Comanches, was an unmarked and crumbling ruin eighteen miles south of Iraan when I lived there. Those who knew where to find the place would still occasionally turn up a uniform button or an abandoned canteen. (It's now a state historical site.) The Goodnight-Loving cattle trail, generally considered to be the worst dry trail drive in the West, reached the Pecos a little north of Iraan at the dreaded Horsehead Crossing, before turning north for Wyoming. After a waterless eighty-mile stretch from the Rio Concho, the trail-weary cattle, smelling the river, would often stampede the last ten miles, according to J. Frank Dobie. "For a day's ride back east from the Pecos," he writes in *The Longhorns*, "cattle ribs and skulls and horns, like sculpturings in chalk, extended the line of Golgotha markers that originally gave Horsehead its name."

Fort Lancaster was also a target of the local gold bugs, who pursued a less-popular but altogether more picturesque recreational enthusiasm than arrowhead hunting. The standard version involved a missing payroll chest, its carrier waylaid by Comanches or bandits, the gold washed away down a creek bed during a flash flood. "Everyone knew" that back in the thirties a bulldozer operator, building a highway bridge, had turned up a gold coin. There were theories, and long discussions, usually around filling station grease racks. And then there was the ancient skeleton someone allegedly found in one of the caves, sitting upright with seven bands of Aztec gold around the bones of its upper arm. The finder had gotten clean away with *his* haul, but there had to be more up there somewhere.

You needed an army surplus jeep, a metal detector, and lots of batteries for your flashlight. You went at night—sometimes the entire night, which is why some of my students fell asleep in class—because if you found something good, you didn't want the whole world watching where you'd gone and where you were coming back from, did you? I was invited but never went along, never really wanted to, although I enjoyed the stories. Nobody ever found anything while I was there, at least that they would admit in public.

My tastes in recreation have changed considerably since those days, but I do look back at the arrowhead hunting as a lost opportunity, wishing I'd done it more. I got sidetracked instead into another mode of exploring the Pecos canyons. Somehow the high school band director, Bob Kern, and I stopped looking for chipped flint and started rolling rocks down canyon walls. Perhaps it was because we were in our early twenties and inconsolably bored with small-town desert life, but I think it had more to do with indulging, more or less harmlessly, the fascination with violence that sometimes infects young men.

I don't remember how we got started. Probably we kicked stones loose as we climbed, by accident, and were amused by the result. Before long we were making specific expeditions, going well back into nameless, unoccupied canyons, carrying along heavy pinch bars to dislodge ever bigger stones. We'd climb up to the rimrock at some spot that offered a good, long downhill slope ending in a gully—dry, of course. We'd find a good-size boulder perched on the edge, preferably one that wouldn't take too much work to pry loose. We'd struggle and sweat and, giggling, finally tip it loose. It would groan, teeter, slowly gain speed. Often it would roll ten feet and stop, more firmly lodged than when it started, and we'd go on to the next—once the stone had lost that initial momentum, there was no chance of moving it again. What we were after was one that went all the way, picking up speed, thundering and bouncing and smashing its way down, knocking loose more scree as it went, bashing through the catclaw and even knocking over the occasional mesquite tree. If it was a good run it would hit the bottom of the wash at maximum speed and go bounding partway up the opposite slope, gradually lose speed, stall, and sometimes then even repeat the whole process in the other direction. We'd sit on the rimrock watching, yelling and cheering it on.

It was as mindlessly stupid an exercise as one can possibly envision, and we became addicted to it. It makes me cringe now—it was purely destructive vandalism wreaked on the natural world—but except for knocking down a few mesquite trees, which locals considered to be a proliferating pest species anyway, I can't see that we were doing much real damage. Speeding up erosion. Still, we felt vaguely guilty at the time, and talked about that, trying to figure out why. Maybe it was because it was such small-boy mischief, so undeniably silly. It reminded me in some strange way of fireworks.

We finally overachieved, with one Volkswagen-size monster that took a good half hour to work loose from its niche in the rimrock. It was so big that if in the first few minutes of struggle we hadn't managed to get a discernible groan out of it, signifying at least the possibility of motion, we'd have gone on to something less ambitious. Groan it did, and we labored on, and got it to wriggle, and then it rocked a bit, and finally we broke it loose. We danced back to safety when it started to move, looking at each other, a little frightened. It was so big we could scarcely bear to watch, but we did, and it was perfectly satisfying, all the way to the bottom. We quit and went home, and never went again, knowing that we didn't want to set loose a bigger stone than that, but knowing also that anything less would be anticlimactic. It was over, an era ended; the next activity that captured my interest was automobile racing.

Now I think I see that rolling rocks was a way of operating the world, of making big things happen, when all other personal power seemed somehow circumscribed, if not by our boring jobs then by that awful—awe-full—geography. High desert country can be glorious. There were special times, fabulous sunsets and dawns, and desert nights that were magical. Most evenings the sky turned an electric blue at twilight and a kind of crystalline serenity settled down over things, no matter what kind of gusty winds had been irritating you all day long. Pump jacks worked away on the cactus-strewn hillside behind our small house, and I remember watching them slowly churn away, silhouetted against that darkening blueness, and thinking it wasn't such a terrible place after all. But you could bear to live there only by ferreting out such moments and fastening on them; the rest of the time, and it was a very large percentage of the time, you tried to hold on to your sense of humor. Mostly it was just very hot and very windy and very, very dry.

Water represents a kind of freedom, I think. I always felt trapped out there, never really felt I was there by choice. In some ways I found a more powerful pleasure than either rolling rocks or looking for arrowheads simply in climbing up the dry creek beds where the bedrock, carved into fantastic potholes and dried-up waterfalls, made me imagine that country re-watered, verdant, as it had to have been at some time eons in the past. In my fantasies I didn't just invent swimming holes, I re-created whole landscapes. If one could only somehow wet down west Texas, I thought, and stop the dust storms, it would be a wonderful place.

I'm also a little embarrassed to remember that I populated some of those imaginary pools with what I could only characterize now as voluptuous nymphs. That usually happened on long drives. I'm certainly not proud of it, but erotic fantasy was generally accepted, even recommended, as one of the more effective methods of staying awake down the endless, hot, long, straight roads, of which in west Texas there are almost no other kind.

12

going outdoors alone

> Where the white pines stood thick, rising one above another . . . and maple limbs were mingled with the pines, I was on the verge of seeing something, but I did not. If I had been alone and had had more leisure, I might have seen something to report.
>
> —Thoreau's Journal, January 25, 1852

When I was younger and a more romantic soul I was a devotee of the symbolic act of plunging into cold water; as I grew older and body heat began to moderate, so did that moment's charm. By middle age I was positively resistant to it, even as a competitive swimmer. I didn't mind the two-hour workouts but would malinger at poolside, resisting the abrupt initial transition from dry to wet.

On our French River trip we slept on solid granite, and despite a good sleeping pad I woke each morning stiff and groaning. Geoff, whose cottage we started from and who had more experience with those chilly waters, claimed that a morning dip, first thing upon awakening, somehow reduced the problem. It didn't seem to me to make physiological sense, but I began trying it—slipping out of my tent at dawn, trying not to shriek at the shock and wake my associates—and found to my surprise that he was right.

Besides, somehow it had gotten pleasurable again. Since then, because we have a nice pond at home, and because in summer when I am not there I try to be on some other attractive water, I've managed to hang on to the discipline. I struggle to relearn it every summer—sometimes after every summer cool spell—but if I push on through, if I go ahead and take the plunge, I am always glad I did. I am generally suspicious of ritual, but a morning dip gives me, in addition to a sharp blast of physical refreshment, a curious existential moment. I dive in and, always, at the transition point of the glide, as momentum dies and my stretched-out body begins curving its way back up toward the surface, I find myself thinking, in effect, oh yes, myself. *Here.* The moment would be more poignant if I did not also sometimes picture a dog bending in place to smell its own other parts.

Solo canoe trips are sometimes like the morning plunge: difficult to launch but worth the stimulation. It's not so much that I really like to go off into the woods by myself as it is that I need it, or think I do—and in fact every time I go I am vividly aware that it is the best thing I've done for myself since the last time. For myself: because the self is what one takes along—the not inconsiderable excess baggage—when one goes alone. Like diving into cold water at dawn, there has never been a time when I wasn't glad I'd done it, afterward if not during.

It is a bright, late-July morning after an evening of crackling thundershowers. Chris has an old friend as house guest, my presence not required. I get gear together and slide the canoe onto the car. The forecast calls for more showers, but the water I'm headed for is half a day to the west, and I hope to drive through the weather. Anyway it'll likely clear up there before it does at home.

The distance and some niggling little errands along the way make it five P.M. before I get to the put-in, and in my haste I choose the wrong entry point, giving myself three-quarters of a mile of extra carry to water. I blame the long drive: if my bladder hadn't been so full when I got there I'd have studied the map harder, and discovered my mistake before I carried the canoe off into the woods. After about twenty minutes of slogging I wedge one end in a tree fork so I won't have to hoist it onto my shoulders again, step off into the bush, and reduce the urgency. Leaving the canoe where it stands, I dogtrot the remaining two hundred yards just to be sure there is in fact a pond over there somewhere. There is. Go back for the canoe, get it to water, head back to the car for the rest of my gear, wondering if there was ever a Keystone Kops routine featuring a solo Kop. Realizing meanwhile that I am an impetuous, precipitate fool, which is the kind of thing you find out when you go alone.

Long Pond is in the St. Regis Canoe Area—no motors allowed, canoeists and backpackers only. It is narrow, riverlike, oxbowish, three miles or so long, with sufficient hidden coves and wandering arms to make sure its several campsites are tucked out of sight of one another. In late July I figure it will be crowded even in midweek, and stop at the first campsite I come to. As I am surveying it another canoe comes by from the opposite direction, reporting no empty campsites thataway. I congratulate myself on my luck and set about making camp, moving at a dogged, methodical pace that precludes both whimsy and introspection: just doing what has to be done, a little more hurriedly than usual before darkness adds to the difficulty. Every part of it is a dutiful, disciplinary task, more or less repetitive, that somehow when you're alone becomes worthwhile for itself: a kind of practice. One doesn't get exalted, there's only a kind of peaceful plodding, a satisfaction. I get tired—I'm plenty tired from the portage anyway—but that's an acceptable part of the deal.

It's nearly seven before I really have a chance to look around; thank goodness for late July's long evenings. Plenty of ripe blueberries are within easy reach for the morning pancakes, if by then I have slowed down sufficiently to collect them. Seagulls carry on a noisy dispute in the distance, whitethroats and thrushes sing more sweetly closer in. Two adult loons and their single chick are on patrol a couple of hundred yards off my shoreline. The parents dive, leaving the nervous-looking youngster alone for a long, suspenseful pause, then simultaneously surface at baby's side, to relief all around, including over here on shore. One of the adults pokes a minnow down the young bird's throat. A beaver glides past my campsite, ten yards offshore, eyeing me suspiciously, then rolls quietly under, minus the expected slap of tail. The water is clear if the usual tannin brown, not terribly inviting, but I am hot and sweaty and at sunset slip in anyway, and as usual am glad I did.

Dusk gives the sky a rosy glow. Canoe-only areas do attract paddling nuts, and the change of light is a signal for most of them to go out for a quiet evening tour. The sight is so beautiful I slide my own boat back into the water too, for a slow circuit of my immediate area, on a mirror-perfect surface, playing with the water, the canoe handling better without the load of gear. Moving easily, gliding so well. Kids' voices float across the water—every campsite but mine seems to have children, dogs, or both. One couple paddles by in a slick tandem, accompanied by a teenage son in his own matching solo canoe. The father paddles in the stern, the mother fishes from the bow with a large spaniel at her elbow, the dog intently watching every twitch of the line. I comment on this; "Oh," the woman says, "he loves to fish." The son also carries a fly rod but seems more interested in paddling circles around his parents, who are eager to talk canoes. "Go closer," the mother says to the son, when I admire his one-seater as he passes. "Show him your boat."

I can't see the nearest campsite but can see its landing, where a ten-year-old boy stands, casting like a pro. Dad and another son are out in the canoe, drifting along the far shoreline. From half a mile away, in the evening quiet, I hear the child softly say, "He got away." My nose is out of joint. Hell, I'd have let my father teach me to fish, too, if he'd ever taken me out on a pond like this, on evenings such as this one.

People are fishing again when I get up the next morning, fishing before the rain that begins to pelt down just as my pancakes come off the fire, fishing for all I know while the rain continues. Fish are good to eat, sure, but this seems excessive. Perhaps I need to reexamine my attitudes about the sport. After all, whether or not Ralph Jerome tried to teach me, I wouldn't have gotten the introduction to the woods that I did without his fishing. I wouldn't be here if it weren't for fishing, whether or not I ever wet a line.

I had neglected to put up a tarp, so the rain makes me bolt for my tent with plate of pancakes in hand, thinking of Ralph, who died when I was fourteen. As well as I can remember he didn't particularly like fish, and wouldn't eat the catch. I never saw evidence that he was interested in how fish thought or what fish wanted. I think he just loved to cast, the sensuous act. He used to give dry-land demonstrations to Boy Scout troops. I suspect he also loved rivers. Maybe the fish part, the creature at the other end of the line, was just an excuse, getting him where he wanted to be: a kind of beard.

Gwen liked to fish. When she lived with us, in her seventies, she'd limp on arthritic knees down to our pond to fish for an hour or so, catching and releasing smallmouth bass. She wasn't obsessive about it, though. Jud would fish if the circumstances were right, but that's not why he went to the woods. Ott never fished that I knew of. Jud, Ott, and Ralph; how strange, almost funny to link them. The men in my life: adult males who somehow kept me feeling immature and unsure of myself longer than I should have. When Jud died, the last of the three, I became the elder, a very peculiar feeling.

Ralph did try to teach me to cast one summer, in the backyard. He rigged up one of his old fly rods for me, with a wisp of cotton on the end of the line in place of a fly. That was all I'd need, he said, to learn to get the line out. He showed me the moves, got the line out easily, made a few demonstration casts, then left for work. I could whip the line through the air all right, in something like casting, but when I made the mistake of reeling in, I couldn't make the line go out again. With eleven-year-old patience I gave it half a dozen tries, to no effect, and then with eleven-year-old logic decided I needed more throw weight. I clipped a clothespin onto the end of the line to give it more heft. My first cast fouled the reel so completely that it took me the rest of the morning to get the backlash untangled so no one would know the sin I'd committed. That evening I suggested we play catch instead.

I've caught a few fish myself, but found the experience unrewarding. I admire some of the gear but have never wanted to own any. I've tried to concoct a moral position in this matter—as with Muhammed Ali and the Viet Cong, I have no quarrel with fish—but I suspect it's just squeamishness. In *Fly Fishing Through the Midlife Crisis*, Howell Raines describes "pulling on a fish" as a source of pure physical pleasure, with "cells scattered throughout the bone marrow—and particularly in the area of the elbows . . . having subtle but prolonged orgasms and sending out little neural whispers about these events." Not for me. I don't mind putting worms on a hook or getting the fish off it, but pulling on the fish is precisely what I don't like: holding on to a line on the other end of which is a thrashing wild thing with a hook in its throat.

The rain withdraws in late morning into low, gray overcast, the day humid and still. I straighten up my campsite and then set out to explore the pond, paddling past other camps where those occupants who aren't out fishing are quietly going about the same housekeeping chores I have just finished, which gives a companionable feeling. The largest of the loon family keeps showing its white breast in the distance, like a flashbulb going off, snatching my attention every time. (Don't you know that when the loon compresses the air out of its feathers and deflates its air sacs and rolls smoothly over into another dive into that deep cool liquid medium—through which it moves like a fish, moves well enough to *catch* fish—don't you just know that that is a moment of pure physical pleasure to the bird? That curl and dive?)

A patch of blue sky shows briefly, but it is clearly what westerners call a sucker hole. I roll the term around in my mind, enjoying the characteristically western attitude that it implies. My interior life, I realize, is mostly putting words on things: writing, really, if only in my head. I find myself composing ahead of time just how I will describe parts of this trip to Chris. This disappoints me. Am I doing it for the trip itself or just to tell about it? Would I do it if there were no one to tell it to?

Okay, so one gets a little silly, soloing. Alone in the woods, you're either trapped in your own head, self-absorbed, or you are absolutely not, dealing with external matters: your attention is either all inside or all out. The trick is to keep it out there, seeing. Keeping your line out. I recently went for a walk in the woods with three other people, including a horticulturist (daughter Kate) and a physical geographer (son-in-law Kurt) as well as another writer (wife Chris). We were hiking through an interesting and unfamiliar place, with much to look at and, with two experts along, plenty to talk about. I am old enough to particularly enjoy being instructed by my children, and we had a fine time, but I realized afterward that we'd walked past things—small waterfalls in trickling brooks, openings onto views, restful glades—that deserved more time. We didn't stop because nobody wanted to hold up the group. At one point I stepped off the trail to get a better look at a wetland, and immediately told the others to go on ahead, I'd catch up. We walked too fast the whole time. A couple would've walked slower, I think, and someone alone more slowly yet.

When you go alone you find out why you are there and what you like and don't like about it in a way that you can't when you are accommodating other consciousnesses. Chris and I prefer to go together, although, or maybe it's because, we share a taste for solitude. There's an oxymoron here somewhere.

As I circumnavigate the pond I keep an eye out for more firewood, which at this time of year can be problematic. There has already been a lot of traffic on the pond, and it is illegal, or anyhow immoral, to cut down any living thing, to gather anything but what has already fallen. Springtime is sometimes worse, when a six-month accumulation of snow has thoroughly soaked any suitable new dead wood. We spent one Raquette River trip trying to use damp paper towels for kindling, which proved to be virtually fireproof. (Now we collect birch bark—infallible kindling—ahead of time.)

In a lovely book called *Maps and Dreams*, about reservation Indians in the interior of British Columbia, Hugh Brody describes an old man's fire making. The elder, named Joseph Patsah, selects each piece of wood with almost maddening care, feeling it, considering it deliberately, assessing its dryness and suitability, almost like a tea ceremony. "He lights a fire in the woods with all the precision and attention to detail that is the untroubled selfconsciousness of those who, with age, are resolved that what they represent be respected," says Brody. Also, he points out, in the subarctic, "fires and walking are the two ways to keep warm—clothes are light—and the old cannot always walk long enough or fast enough to keep out the cold. Excellence with fires is essential. Joseph Patsah does not flaunt his skill."

Most people play with the campfire when they're in the woods. I certainly do, I can't help it. Chris says it's a male thing, men as fire tenders, but she can't resist messing around with the fire either. One is compelled to keep trying to make it burn a little better: feeding twigs, pushing the burning brands about, getting fuel and flame to the right place. It's not exactly a pointless exercise, since inattention can delay the spaghetti for an hour—or, see Joseph Patsah, worse.

Anyway, the fire tender can't regard the fire as private turf, even if the member of the group who gets the first fire of the day going is something of a hero. Everyone else emerges on his or her own schedule, stretches, yawns, goes to pee, then comes over and starts messing with the fire. It's what you do. When you're solo you have the fire all to yourself.

I find an abandoned beaver lodge with a stock of driftwood piled on top, and load up the canoe with enough to take me through tomorrow's breakfast and leave some for the next arrival. When I get back, the fellow in the next campsite has stopped fishing and is chopping away at something with an axe. The sound goes on and on, through lunch, through an otherwise peaceful afternoon snooze. Every once in a while I hear what sounds like a tree crashing to earth (and once a child's cry: "Dad? Dad? Are you all *right*?"). It baffles me. Chopping things down is simply not done, not allowed. The sound carries all over the pond. I assume someone will eventually go talk to the guy, explain the rules. I know people who would go do that immediately, but I am not among them. After a couple of nervous encounters with armed hunters in no-hunting zones, I have sworn off chastising hot, sweaty males with weaponlike objects in their hands. So I slip the hawser of responsibility. That's on top of being impetuous and precipitate. Not everything you learn about yourself while soloing is pleasant news.

For my own evening fire I have lucked into a pine knot, which burns slowly and cheerfully for hours, providing a kind of pilot light when I want more flame. As a kid I'd read about pine knots but had never burned one before, never had the wit to look for one specifically. I don't know woods by their species and burning qualities, as woodspersons should. I had no mentor in that regard, no Joseph Patsah, and have not paid sufficient attention as I've gained experience. There's seldom enough firewood around to allow choosiness anyway.

Jud's woodcraft was even more helter-skelter than mine. I could have had other mentors—there were volunteers, scoutmasters, other kids' dads—but I never found one I would listen to. They were always treating me like a kid, and laying down romantic, high-blown notions about what it meant to go into the woods. They gave me no confidence they knew what they were talking about—except for Coleman Ward, and I didn't get enough time with him.

I was too young at the time to judge by result. Ralph was a master with the casting rod, Ott could make airtight, solid joints, Jud spun stories to make what we were doing more interesting than it perhaps was, or seemed. None of them ever called my attention to the thing itself, or expressed any pleasure in it. I don't recall an identifiable sense of pleasure in any of them. In memory, anyway, I always had a better time than they did, for all my whining.

I always felt I was supposed to take up fly-fishing, or at least felt that Ralph would have been pleased if I had, but I didn't like the associations it had for me and so never did. When I say that he was an alcoholic and a fly fisherman in that order, the order is mine of course, and I am surprised once more by the anger in it. I keep thinking I'd gone beyond all that. I've always characterized him that way, almost from the moment he died. I once wrote elsewhere about his drinking, and a reader sweetly asked if I realized I was one of those adult children of alcoholics. I'd talked about having a sense of unfocused urgency, of time pressure, which my reader pointed out was characteristic of ACOAs. She called her own similar uneasiness "the churning," which sounds about right. I have noticed that in the woods the churning stops for me. I suspect that when Ralph picked up a fishing rod his churning stopped too.

He didn't live long enough to find out whether I became a fisherman or not. I was not cheated of anything but a certain amount of his soberness, and the opportunity to know him as an adult. I am now twenty years older than he was when he died, and am startled to consider that he never grew old. He died so young that he never really even had a glimpse of what it was like to age.

Jud picked up Ralph's storytelling bent and improved on it, tipping over into the realm of the whopper. He became a poet, often preferring narrative poetry to lyric. I became a writer too, if of a more pedestrian stripe, but I've never been able to resist that little filigree of exaggeration that makes a story more dramatic. I'm not sure I actually saw Teacup come out of that pond with the fish flapping from his arm. I remember clearly the sight of the hook in his flesh, but the fish may have been removed by the time I got there. I have replayed it in memory so many times I can no longer remember what I saw, what I heard, what I imagined.

Good storytelling should seek out the river within the river, the zone of fastest flow. My apologies for the eddies in mine—but you wouldn't want one of those Army Corps of Engineers rivers, would you?

• • •

The problem is that our attention is seldom where it ought to be, and in those very few moments when it does get to where it ought to be it demonstrates to us—in a flash of recognition, a burst of pleasure—how misplaced it usually is.

I suppose that's what practice is about, the practice of the canoe stroke, the practice of making fire. Mystics speak of the practice of the Tao. I don't consider myself a Taoist, but I often play a bedtime game—particularly sleeping outdoors, when the bed has a certain ineluctable firmness—in which I sense gravity not as a great force pulling me down but pushing upward against my body, pushing with 170-odd pounds of pressure directly from the center of the earth. I amuse myself turning all of physics upside down that way: the sight comes out of my eyes and grabs everything upon which light is allowed to play. Perhaps it is the Tao that shines sight out of my eyes, the Tao that makes the nerve endings in my retina shine out and create the visual world, the Tao that operates the nerve endings in my ears to fill the woods with birdsong. Maybe all of the birds come out of the Tao. Nature loves me this I know for the forest tells me so.

At sunset a great blue heron comes wafting into the cove across from my campsite, reminding me to watch the fliers. In the still air the gulls put on the best show, their glide patterns beautiful, with wonderfully complex landing patterns and big flare-outs. Unlike the lumbering loons, they take off directly from the water, straight up, with a single wing beat.

So do some ducks. They must have much less wing loading than the other waterfowl.

I lived for a time in a suburb of Denver, at a time when we were barely making ends meet. I loved the mountains that loomed over us, and went up into them every chance I got, spending money we couldn't spare for the gas to get there. When I was there I didn't have anything to *do*, which doubled my frustration. I wanted to ski but couldn't afford it, certainly couldn't afford to take the kids. I rock climbed a little, for the athletic fun of it, but that was irresponsible for a family man, particularly unequipped, and that equipment was expensive too.

When my fortunes improved I did ski, by then in the East, and found other appliances to justify going to the woods, but over time the need for those appliances, and the attitudes that went with them, began gradually dropping away. Maybe it was just age, although I don't see many old men giving up fishing. Wouldn't that be a sight, though, the lakes of the world full of old men, sitting out there in their boats without fishing tackle in their hands?

"I have been surprised to consider that the only obvious employment, except wood-chopping, ice-cutting, or the like business, which ever to my knowledge detained at Walden Pond for a whole half day any of my fellow-citizens . . . was fishing," says Thoreau in the chapter entitled "Higher Laws." "Commonly they did not think that they were lucky, or well paid for their time, unless they got a long string of fish, though they had the opportunity of seeing the pond all the while. They might go there a thousand times before the sediment of fishing would sink to the bottom and leave their purpose pure; but no doubt such a clarifying process would be going on all the while. The governor and his council faintly remember the pond, for they went a-fishing there when they were boys; but now they are too old and dignified to go a-fishing, and so they know it no more forever. Yet even they expect to go to heaven at last. If the legislature regards it, it is chiefly to regulate the number of hooks to be used there; but they know nothing about the hook of hooks with which to angle for the pond itself, impaling the legislature for a bait. Thus, even in civilized communities, the embryo man passes through the hunter stage of development."

13

the geography of dreams

> When I asked people why they liked where they lived they always talked about the landscape, and then how near they were to somewhere else.
>
> —Michael Parfit, *Chasing the Glory*

Recently I had one of my recurring escape-from-Manhattan dreams. I was without a car, and getting home was somehow more complicated than I could manage. When I asked for directions in the subways, I couldn't understand the garbled answers. Taxis ignored me. I walked for long distances through the city, worrying at the problem. I couldn't get off the island. Manhattan seems to represent the ultimate Non-Home in my dreams, even though I've lived there twice. The home I am trying to get to, in these dreams, is always to the north. My only other dreams in which geography plays a part, in which there is geography at all, have to do with that home up north too.

When I last lived in New York I joined a health club with a swimming pool. I was recently divorced, I'd just started a demanding magazine job, everything in my life was new and unfamiliar. I was a little disoriented, and groping, I suppose, for connection to something outside the citiness of the place. I swam at the health club only four or five times, though. It was not a pleasant place—tepid water, not terribly clean, something sinister and entirely too sexualized about the dressing rooms—and I quit going. No more swimming, I said at the time; I was sick of having water in my ears.

It was my second go-round at New York, this time with a hotshot job, an apartment in the Village, the riches of the city before me—and I couldn't wait to get out of there. I don't have towering ambitions, but I wouldn't mind being recognized as having been perhaps the worst New Yorker in the history of the city.

Eventually I remarried, with astounding good fortune, and did escape. I'd taken up skiing, and for that and other reasons inveigled Chris into moving with me to northern New Hampshire. It was home for me the moment I got there, in a way that no place else I'd lived, including Texas or Oklahoma, had ever been. New England is my psychic home, I can't explain why.

Geographically, it's hard to think of Manhattan as much more than a plat map surrounded by murky waters; northern New Hampshire, by contrast, was the real McCoy: physical geography writ large. The place we moved to was on the western edge of the White Mountains, where they start off with a bang, rising abruptly and spectacularly some thirty-five hundred feet above a gentle north-and-south valley. It's called the Kinsman Ridge, for Nathan Kinsman, who settled below it in the 1700s and gave his name to one of the peaks along its length. The valley below it should be called the Ham Valley, in honor of the Ham Branch of the Gale River that runs through it, or the Franconia Valley, for the larger town that sits at its north end, but it is known locally as the Easton Valley, for the tiny hamlet of Easton, about halfway down its length. We were to live there for twelve years—from, essentially, early adulthood to early middle age. As we used to say in the sixties, all our changes were there.

We were awed by our new surroundings, and at first a little frightened. Chris, whose upbringing was less rural than mine and who wasn't nearly as interested in skiing, now admits she thought moving there was crazy. One night in our first winter the temperature dropped to forty-two below. (I'd brought the battery in the house overnight, so the car did start the next morning, but when I sat down in it, the plastic seat covers shattered.) Winter went on forever, followed immediately by blackflies, which filled us with despair. We remained tentative about the place for a year or two, and used magazine assignments to scout the mountain west and Pacific coast for a place to move to, but we couldn't find one we liked better. After New Hampshire, everywhere else seemed too crowded.

We also never found a way to use those New Hampshire mountains as fully as they deserved, perhaps because living there full time was a powerful enough experience in itself. As they say in real estate, location, location: the physical fact of where we were became a dominant part of our lives. It took us a few years to realize what a good thing that was.

When we did use the mountains it always had to do with water more than rock. Skiing was only the most obvious way, and we found ourselves doing much less of that over the years. We hiked when weather, bugs, and energy levels were in proper conjunction, preferably following streambeds. Clear water running over smooth granite is high on my list of the natural world's most satisfactory phenomena, and New Hampshire has as much of that as anyplace. We knew where one small stream tumbled down sheets of granite for miles, over falls and into and out of potholes, carving out slides and chutes as curvily voluptuous as Henry Moore sculptures. They cried out for you to curve your body into them—if you could stand the cold.

What kept us from getting blasé about our local mountains were summer visits from my kids, devoted primarily to the search for new swimming holes. This replication of the postdivorce summer visits that Jud and I had made to Oklahoma was unintentional but fated: what else do you do in summertime when kids come up to visit their dad and new stepmother? One summer Gwen came along, by then in her sixties, and we got her plunging into icy mountain streams too, even if a bit of the spring had gone out of her plunging. ("Icy" is not entirely hyperbolic—we once found a narrow, shaded notch on the slopes of Mount Kinsman that still had ice on its walls in July.)

We were never very systematic in our search for swimming holes, although we did ask around. More often than not we'd just get in the car and head for some mountain road, stop wherever water crossed, and scout uphill and down in hopes of finding a nice pool, waterfall, or other piece of streambed that looked fun to play in. We found enough suitable spots to become choosy about water quality. Even in northern New Hampshire, streams that start out glass clear in the mountains darken quickly when they reach lower elevations and begin slowing down. We did most of our exploring on the uphill side of the road.

Unfortunately, the higher you go, the colder the water, even in August. The best swimming hole we ever found—the only one on my life-list that I'd rate better than the dam on the Comal in Texas—fit that category. It is on the southwest slope of Mount Washington, a mile or two up the road from the gaunt old hotel that was the site of the famous Bretton Woods Conference of 1944, which created the International Monetary Fund. The swimming hole is, formally, the Upper Falls of the Lower Ammonoosuc River. Its gloriously clear waters drop first over a twelve-foot waterfall—which you can use as a water slide, or hide behind—into a succession of enormous potholes, the first two of which have a connecting underwater passageway big enough to swim through. On three sides of the potholes are cliffs to dive from, your choice of ten or twenty or thirty feet high; the potholes below are plenty deep enough. It is God's own best idea of everything a swimmer might possibly want. Unfortunately, the water never gets warm enough for extended human use.

Normal humans, anyway; the local hotbloods seemed to be able to stand frequent if not continuous immersion. Some of them liked to swan dive from the highest point on the cliffs into the smallest pothole, which would comfortably hold twenty swimmers at a time but from the lip thirty feet above looked about the size of a washtub. Their dives terrified me, doing to me just what I so used to love doing to visiting tourists back in Texas. If I had grown up around there I'd have had to do that too. It is so nice no longer to need to.

The Easton Valley is regarded locally as a particularly beautiful and pleasant place; it and its overlooking ridge form a natural unit, one you can get your mind around. I suppose New England is full of such places, tidy little enclaves that you can come to know with a fair degree of intimacy—although "intimacy" is a strange term to use about any aspect of New Hampshire. The old-timers certainly have a feel for every dip and hollow, every turn of the breeze that blows their length and breadth. They can tell you what to expect, if you can figure out how to ask. Lives are formed by these realities of terrain, small societies created, one way or another, around their exigencies.

The exigencies are stark, and clearly understood. Easton, for example, is at about 44° north latitude. Manhattan is at 40°45′. For every degree of latitude you move north, winter comes four days earlier and spring four days later. A degree of latitude is roughly equivalent to four hundred feet of altitude. Easton isn't very high, at eleven hundred feet, but nearby peaks reach five thousand feet. I make it fifty-two more days of winter than in sunny Manhattan—that's climatological winter, as opposed to the astronomical one—and up on the ridge, twenty more days on top of that.

Mountains also generate their own weather, being colder and wetter on the prevailing upwind side, warmer and drier on the downwind side. (Easton was upwind.) The valleys below get less sunlight, and collect the dense, cold air that slides down the surrounding hillsides. One grew peas in Easton but not ripe tomatoes. Mountains are also consistently windier than level country. The windiness comes not only from the great temperature variations between exposed peak and shadowed valley, but from compression: prevailing winds get squeezed tight as they are driven upslope, and, thanks to the Bernoulli effect, pick up speed.

There's an Easton Valley wind that demonstrates this

principle so well it has its own name, the Bungay Jar. Some say this is because the south end of the valley used to be called Bungay Corner—nobody knows quite why—and that's where the wind seems to start. Jar is what it does, particularly in early spring, when it builds to a strange periodicity, with eerie calms between sharp jolts and blasts. Sometimes the valley bottom will be relatively calm while great winds are roaring up on the side of Kinsman; then, the locals will say, you can hear Bungay talking. When the Bungay starts up, the first four or five blasts will make you put down your book and go to the window to see what the hell is going on. Hell is going on. You get used to it. The highest wind speed ever recorded on earth was just a couple of dozen miles farther north, on the summit of Mount Washington. The Bungay Jar has to go somewhere.

It was not always hell going on outside, however, although I do remember setting out one Christmas afternoon to try to buy milk in minus-thirty-degree weather. It was one of those dreadful open winters, with no snow cover to soften things, nothing but iron-bound ice over fractured granite, and as I drove from shuttered store to shuttered store in the early dark, truly worried about the possibility of car trouble, I remember imagining that along with the scraps of paper, ice dust, and powdered dog shit that a vicious north wind was blowing across the road in front of me, I could make out some dim something I decided must be the ghosts of dead bears. Conditions like that do stir the imagination.

We liked to romanticize things anyway, particularly for outsiders. One summer day an editor in New York phoned for Chris. She's gone berrying, I said, she'll call back. Afterward I began picturing the image that editor must have conjured up—Chris strolling through a dappled meadow in billowy summer frock, a sun hat, perhaps with trailing ribbons, a wicker basket on her arm. An hour later Chris came in the door cursing the goddamn deerflies, her gimme cap filthy with bug dope and sweat, plastic bucket still looped through the belt of her overalls, bleeding scratches from fingertips to armpits from fighting her way into and back out of the gravel pit where the best raspberries grew. How'd it go? I asked. See any bears this time?

The swimming we were doing—in shallow creeks and potholes, in the few weeks of summer when water temperature and insects allowed—was inconsequential, just a way of entertaining the kids. We were not so interested in total immersion as we were in the beauty of the setting, in finding more and more jewellike little places in which to splash. I had no impulse to do more. After we'd lived there for about ten years, however, and I'd more or less forgotten what it was I used to love about the sport, we were invited one summer evening to a lakeside cabin for dinner. We'd been advised to bring swimsuits and were encouraged to go for a dip before dinner. Neither our hosts nor Chris felt like it at the time, so I went in by myself, just as the sun was touching the horizon, and swam a long, slow way up its brilliant yellow path, dazzled and blinded and almost weeping with the resumption of that familiar pleasure. I came out of the water stunned.

I must have brooded about it over the winter, because the next summer, for what I maintained at the time were reasons of physical fitness, I began going to Echo Lake in Franconia Notch, about six miles from our house. It's as beautiful a spot as you'll find in New England, a tranquil little pond, really, maybe a mile long, with dramatic great overhanging cliffs.

Bierstadt painted it in 1861; the Old Man, the granite profile that is New Hampshire's state emblem, juts out from another cliff just a bit down the notch.

There's a state-run beach at one end of the lake that offered very nice swimming, but it tended to be crowded, so I'd sneak in very early, before any lifeguards showed up to stop me. I'd swim a slow mile or so, making laps just outside the ropes of the public swimming area, trying to get back some feel for the water. It gradually returned, and that too was a kind of homecoming. I had no idea how much I'd missed that long, languorous, rhythmic drive, the chance to work the water. Absent the adolescent distractions, I realized, that was all I'd ever really wanted from swimming. That and cooling off—and unfortunately, I found it out in a place where cooling off wasn't required all that often.

Something about northern New Hampshire, the looming physical fact of the place, produced specific changes in the two of us. It pulled us out of doors. Chris became a gardener, in more sense than one getting back to her roots: rhapsodic summers of her childhood had been spent on an Ontario farm where an aunt and uncle taught her the rudiments of growing things. I became something of a fitness nut, as a midlife response to the natural deterioration of aging.

Both of these activities were severely limited by the climate, but we were about ready to come in out of the cold anyway. Nothing was holding us in New Hampshire (which Chris now characterizes as the rock *and* the hard place) but friendships and stubbornness. We found a house in western Massachusetts that came with plenty of garden space and a fifty-meter pond. I began middle-aged swim racing, and

Chris, after twelve years of gardening without a single ripe tomato, came into the house one summer day announcing that she'd just picked 104 of the red beauties. (The next day she picked 108, and we suddenly had on hand several years' supply of tomato sauce.)

After a year or two in Massachusetts we began making an annual fall trip back to Easton, entirely out of nostalgia. One warm September night we parked on the road near our old house and went for a moonlit walk around a familiar loop, the same walk we had taken almost daily during the twelve years we'd lived there. As we strolled I realized with a small shock that even in the dark I knew every damp spot and bump in the road, knew that patch of terrain as I knew no other on earth. That's the moment we first realized that we had lived in the house in Easton longer than either of us had lived at any other single address; that it was, in some essential way we'd never thought about before, our true home.

That's also when I learned that the Easton Valley is the geography of my dreams. In the years since we moved away I have dreamed of our house there fairly frequently. Sometimes it has changed, with whole new ells and wings, or extra barns and outbuildings on the property. Sometimes the house is moved up or down the road, but it is always on the east side, with a stream behind it, and the setting is always the same. It is in a gentle, north-south valley. There is a wall of mountains, solid and reassuring, to the east. Even in dreams I know that at the mouth of our valley is a larger town and to the south something sinister is blocked off. Most compelling of all is the sure sense, in these dreams, that to the north of the next village there is . . . nothing. The North: wildness, mystery, solitude.

I still find it shocking to have fetched up for so long in

such a wintry place, when I'd always been summer-crazed. How could home be north when I am a southerner? It has always been north, though; fleeing home and trying to get back home at the same time. Chris associates the South with a certain predilection for guns and ropes, but it wasn't the incipient violence that drove me out. I hated the South's racism—and its institutionalized reverence for ignorance—every moment I lived there, but what really drove me out was the geography itself, particularly the coastal plain: the flat, turgid, hot, drooping physical exhaustion of the place.

It is odd to speak of intensity of geography, but that's the strongest realization New Hampshire gave us: that physical geography can be intense, the more intense the better. And you have to go out into it, immerse yourself in it, to appreciate that intensity.

The intensity is always raised by the presence of water. Set and setting, as we also used to say in the sixties, are everything. If the immersion doesn't seem to be working, you haven't gone deep enough. You have to remove the barriers between you and it (car windshields, camper trailers and RVs, motels, even primitive cabins, I suspect). Northern New Hampshire was the most barrier-free place we'd ever lived.

I don't mean to propose that my dreams are of any great significance, but when I mentioned this to Chris she realized that the geography of the place was in her dreams too. That's the haunting thing. I suspect that in the dreams of everyone else who has lived for long in that valley, the world is arranged in pretty much the same way. Perhaps that's the true definition of a countryman: someone whose dreams have the same geography as your own.

14

diving into champagne

> Water is very special to Canadians. There is no Indian word for wilderness, because while we may regard it as something separate from us, for them the wilderness is everything—their dwelling place and source of food, part of their being. To some degree, Canadians view water with that kind of internal attachment.
>
> —Geographer Derrick Sewell,
> quoted by Peter Newman
> in *Caesars of the Wilderness: Company of Adventurers*, vol. 2

Perhaps the single most famous Canadian of the 1930s was a middle-aged author named Grey Owl, an eloquent spokesman and campaigner for wildlife conservation and native rights. He wrote both adult and children's books, made films, shared his cabin in a national park with a family of beavers (and kept a pet moose), and delighted lecture audiences all over Canada, England, and the United States. In person he was Amerindian personified, a tall, grave, almost forbidding man who wore his long black hair in braids, a mesmerizing speaker who lectured in full native regalia. His promotional material billed him as "a modern Hiawatha," the half-breed son of an Apache woman and a Scottish frontier scout. Nobody did more than Grey Owl, in the early days of the conservation movement, to focus attention on the plight of both the wildlife and the native peoples of Canada and the United States.

In fact Grey Owl was an Englishman named Archie Belaney without a drop of native blood in his veins, who dyed both hair and skin to pass as Indian. (He explained his blue eyes as a legacy from his white father; that much was accurate.) He'd immigrated to Toronto when he was eighteen, followed a mining boom to the Temagami region in northern Ontario in 1906, and begun hanging out with the local Ojibwa, attempting to learn trapping. He did become a skilled canoeist and seems to have had a real affinity for wild creatures, but other wilderness skills eluded him.

He was also something of a scoundrel, a binge-drinking physical coward who serially married and abandoned, usually pregnant, an Ojibwa, a Métis, an Englishwoman, an Iroquois, and a French Canadian, without ever bothering to get a divorce—although the Englishwoman, the only one he did not succeed in impregnating, did divorce him.

Scalawags have their own peculiar charm, but my personal interest in Belaney is geographic: his masquerade had its birth in the Temagami region, a six-thousand-square-mile district two hundred and fifty miles north of Toronto that was set aside in 1901 as forest reserve. Lake Temagami is one of the largest in northern Ontario, but the region also contains half a dozen other lakes of sufficient size to be named on highway maps, and several hundred—or thousand—more that are too small to be identified but each of which is exquisite in its own way. In Archie Belaney's day the Temagami region must have been a paradise; in some ways it still is.

• • •

Gwen's sister Orpha once visited Canada and came back to Texas raving about the beautiful lakes. To swim in them was like diving into champagne, she said. The image thrilled me at the time—I must have been about fourteen—but I didn't remember it until I'd found some Canadian lakes of my own, and gone back several times.

My lakes were in the Temagami region, and I didn't find them for myself, I was shown them, an outright gift. It was the summer of the rental canoes, a month before Chris and I were to make our long Adirondack trip. Jo and Mike organized it, a sixty-odd-mile loop through several lakes, over seven days. Mike had made one such excursion before, with Jo the previous summer; Jo, an experienced tripper and a graduate of Outward Bound programs, knew the region well. Another experienced Canadian couple, Doug and Mary Lou Brock, joined us.

The town of Temagami is a five-hour drive from Toronto, and we didn't get onto the water until midafternoon, putting in at the end of a miserable ten miles of corrugated gravel road. Chris and I were nervous neophytes, trying desperately to be cool and casual and above all not to hold anybody else up, but I was more than a little distracted on other grounds: we launched from a heavily used public dock where speedboats jostled for space as we loaded, but I kept looking down alongside our rental canoe into extremely clear blue-green water, the bottom visible ten feet down. It was a hot afternoon—much of central North America was sweltering under a record heat wave—and I couldn't recall ever seeing such inviting water.

We got under way dry, however, on the implied promise of

better swimming later, and proceeded to paddle longer and harder than I'd imagined possible, down a cottage-studded arm of Lake Temagami to a little fifty-yard lift-over into a less-crowded bay, and then through alternating bays and narrows on and on into the late-afternoon sun. Chris and I brought up the rear, studying our fellow canoeists ahead for clues about how to behave. It had not yet sunk in that one paddled in synchrony with one's partner—our small solo canoes at home used double-ended paddles, set your own pace—and I remember noticing how the others' paddles would flash simultaneously in the sun. I began timing my strokes in the stern to Chris's in the bow, and the paddling immediately became easier for both.

It was July in a high latitude, the sun would not set until quite late, there was plenty of time. What Chris remembers most about that day was the sight of her first loons, cementing a personal relationship with that species that has never flagged. (The next morning one would calmly surface within ten feet of her while she was swimming.) We finally camped on a low granite point with a stone shelf sloping gently down into the water. Everyone went for a quick dip to cool off, then we made dinner and fell in bed. I was so excited I slept badly in spite of my fatigue, and woke, sweating on top of my sleeping bag, at about one in the morning. I slipped down to water's edge, where for the first time in my life I noticed stars reflected on a still water surface, the bowl of night above repeated below. I slid on down into the water and sat still on the granite ledge, chin-deep, absorbing the chill, thrilled virtually into a trance simply with where I was. After awhile I went back to bed for a few more restless hours, but about a quarter to five, with light just beginning to show in the east, I got up again—clothed this time—and, with a borrowed fishing rod, paddled out into the middle of that immense bowl of fading night.

By then there was a foot of mist hanging over the water, dousing the reflection of the stars and absorbing the splash of my lure so that I couldn't really tell where I was casting. I had no desire to catch a fish anyway, and kept casting only as a rationale for being out there—going through the motions with less than half a mind, much more interested in looking and listening as the sun came up. One of the sounds I listened to, in my imagination, was Ralph Jerome spinning in his grave. I was certain he'd never managed in all his fishing life to get himself to a place nearly so beautiful as this. And there I sat, not giving a tinker's dam whether I fished or not, in a spot for which he'd have sold his soul. The thought had me laughing around a lump in my throat.

When Chris and I began these excursions into the wild, I hadn't realized how they would plunge me back into those old blood relationships, would put me to pondering family. Partly it was from the way that water and memory stir each other, I suppose; partly it was because as I relearned to perform familiar but long abandoned small labors, I kept finding their hands attached to my wrists.

Belaney's interest in Native Americans seems to have been perfectly genuine, and was certainly lifelong. Throughout a rather lonely English childhood he'd played at being a Red Indian, reading about them and collecting anything Indian he could get his hands on. In the Temagami region, however, no one would hire a greenhorn with no skills as a guide, or grubstake him as a trapper. He worked odd jobs at a local inn instead, where he met, impregnated, and married a kitchen helper named Angele Egwuna. The Egwuna family liked him, and invited him to trap with them in the winter of 1909–10. He later characterized this introduction as his official adoption by the Ojibwa.

He seems to have been an indifferent woodsman, however: "This thing of hunting and living in the bush generally is not what it is in books," he wrote in a letter to the English aunts who had raised him. "It looks very picturesque and romantic to wear moccasins, run rapids, and shoot deer and moose, but it is not near as interesting as it seems, to be eaten by day and night by black ants, flies and mosquitoes, to get soaked up with rain, or burnt up with heat, to draw your own toboggan on snowshoes and to sleep out in sixty or seventy degrees below zero." He did study the Ojibwa language earnestly, and found work for a couple of summers as a counselor at Camp Keewaydin on Lake Temagami, where he instructed wealthy American boys in much the same Indian fantasy life he'd invented for his own English boyhood. When a daughter was born to his Ojibwa wife, he fled to Toronto and made his first acquaintance with alcohol. He would revisit Temagami fourteen years later—after marrying two other women and siring at least one son—and visit Angele one more time, get her pregnant again, and leave, never further to lay eyes on her or their two daughters. He was done with the region: except for service in World War I in Europe, where he was wounded, he lived for the rest of his life elsewhere in Canada, but always in or near the bush. ("Bush" is Canadian for any unsettled region, bushy or not. It's a handy term, much more useful than our "wilderness," which has become subject to hair-splitting legalistic definitions ever since we began trying to save it.)

Temagami didn't miss Belaney. The region was discovered by sportsmen as well as miners and trappers, and came to play a large part in the birth of two North American social movements. The first of these was "summering," or urban flight for recreational purposes, which exploded in the latter part of the nineteenth century and led directly to the development—and, some would say, despoliation—of the Adirondacks, among other sublime wildernesses. The second was the youth camp movement of the 1920s. Temagami had two of the most famous, Keewaydin, with a mostly American clientele, and Camp Temagami, serving primarily the scions of wealthy Canadian families. Both camps claimed to be building the character of future leaders of "the two North American democracies"; both put heavy emphasis on canoe tripping, which would become a hallmark of Canadian life.

The voyageur tradition had already placed the canoe firmly in the national consciousness. Because of Canada's watery geography, the canoe held its own as a means of practical transportation well into the eras of rail and aviation, and in some senses still does. Meanwhile the Group of Seven, an informal band of young artists who broke with European Impressionism to create the first Canadian school of painting, used canoes to explore the Precambrian Shield country, capture it on brilliant canvases, and present its wild beauty to the Canadian people. Canoe tripping was the most practical, not to mention romantic, way of going to see it for yourself.

Most visitors approach Temagami straight north from Toronto, or come up the Ottawa River valley northwest from Ottawa; in either case you climb up onto the Shield, over some of the oldest surface rock on earth. You see it everywhere in highway cuts, exposed granite of a characteristic pink or reddish brown. (Ontario highway crews sweetly attempt to match its colors when they paint out graffiti.) Most of the Shield country is rolling plain with low, well-worn hills and an occasional more rugged ridge running northwest to southeast, at the angle along which the icecap withdrew. A little west and north of Lake Temagami is Ishpatina Ridge, at 2,275 feet the highest point in Ontario. It's not a startling amount of relief, since the lakes themselves are nearly a thousand feet above sea level. It is high country, but you aren't exactly in the mountains.

The region sits in what was, before extensive logging, a red pine and white pine belt, and those trees are still the visually dominant species. Red pine is a very nice tree under which to camp; white pines make a great skyline. Logging controversies continue, as the region is still, for environmental as well as recreational purposes, inadequately protected. Metals were important before the timber was, in the race to turn empty real estate into cash. Silver deposits brought a boom to the area in 1900; the next town east is named Cobalt, and there was once a copper mine on an island in Lake Temagami itself. That eventually resulted in the ten-mile access road that we used to reach the lake, and brought hydroelectric power to a few of the central islands. You also are not exactly in wilderness. One unforeseen side effect of the access road: recreational visitors began skipping the village of Temagami, to the chagrin of local merchants.

Temagami means "deep water near shore," which it has in abundance. The main lake has fifty thousand surface acres, with over twelve hundred islands and five hundred miles of shoreline. There are quite a number of cottages on

Temagami and the surrounding lakes, many of them on islands, and there are fly-in fishing lodges on the more remote waters. Cottages are an irritation to canoe trippers (and, because of vandalism and bad manners, vice versa), but in the Temagami region there's no justification for complaints of wilderness violated: bush it still is, but the region has been inhabited for five thousand years. There are portages that have been used so hard for so long they've been worn into deep, U-shaped troughs that look as if they were dug by a miniglacier. The region is the aboriginal homeland of the Teme-augama Anishnabai—originally about four thousand square miles, of which the tribe has managed to hang on to about seven hundred acres.

Our introduction to the traditional Canadian canoe trip was a seven-day expedition that covered thirteen lakes. I remember it entirely as golden light, miraculous water, and aching fatigue. Successive fifteen-mile days didn't seem to bother the rest of the party—twenty- and thirty-mile days are unexceptional for serious canoe trippers—but wore Chris and me to a frazzle. Taking advantage of the northern latitude, we sometimes paddled until 7:30 P.M., which meant setting up camp and making the evening meal in a haze of weariness. Tired or not, we had a nearly perfect trip, with bright hot days (and frequent immersions), substantial food, a congenial group. Mornings were best, setting out on what seemed like a new lake every day, the underlying tint of the clear waters changing subtly from lake to lake. The few connecting rivers and creeks were better paddling than the lakes, full of lily pads, reeds, beaver dams, and birds, the biota virtually in our laps. The portages were hard work but not brutal. Chris and I came away knowing we'd found a new passion, and set about improving our practice of it. We wanted to go deeper.

Grey Owl destroyed his health with drink (as did Ralph Jerome) and died early, far from Temagami. (From photos there even seems also to have been a physical resemblance between Grey Owl and my father. There are perhaps other than geographical reasons why Archie Belaney captured my attention.) His minor Temagami scandals were soon enough supplanted by juicier stories from the region. The director of Camp Keewaydin, who had a proper wife back home in Philadelphia, chose to soften his Temagami nights with the companionship of a half-breed woman. Told by the board of directors to desist, he refused, was forced to resign, and, declaring himself unable to continue without "the flutter across his face of Temagami's 'keewadin' (northwest wind)," committed suicide, according to *The Temagami Experience*, by Bruce W. Hodgins and Jamie Benidickson. (Much of my information about the physical geography and history of the region comes from the same source.)

Youth camps are small, necessarily flawed institutions that, like most outdoor activities, inspire devotion and invite satire. Those of the Temagami region were no exception. Keewaydin Camps, Ltd., was so successful in the twenties that it became a conglomerate, with satellite camps in Montana, Ohio, Vermont, and England, and a corporate office in Philadelphia. Staff members worked harder in the off-season than they did in session, recruiting at U.S. prep schools on a commission basis. At least during the off-season they had some control over their own lives; Keewaydin was all male by executive fiat, staff wives and daughters sequestered at the other end of the island from the camp. No visiting was allowed, no drinking, and no fraternizing among cottages elsewhere either, regulations honored primarily in their breach—and bound to foment rebellion. When the Depression began to shrink enrollments, the new head of the operation cut out the commissions on enrollment. Three staff members, who had in the past personally signed up more than half of the campers, opened a competing camp, called Wabun (for the east wind, or dawn), in 1933. Both camps survived and eventually expanded—one used green canoes, the other red ones—and became fixtures on the Temagami lakes. Both were dedicated to wilderness canoe trips, some of which went as far north as James Bay.

Chris and I were eager to go back to Temagami, and did so, once with Mike and Jo, the year after that with another couple, seeing a new assortment of lakes each time. (They are all gems. The special ones are not advertised; you get topo maps and local information, and go from here to there and then there, enjoying the ride.) Last year we went by ourselves, a trip we sneaked in late, after we'd pretty well given up on canoeing plans for the summer. We were in a hooky-playing frame of mind, off on a lark, which made even the six-hundred-mile drive a pleasure: up onto and across the minidome of the Adirondacks and then into Canada, where the houses are smaller, the signage subtly different, the agriculture distinct enough that even the crops in the fields have unfamiliar hues. Up the Ottawa Valley, teased with glimpses of that mighty river, and onto the Shield itself, pink granite everywhere, white pine and red pine forests closing in on the road. It is still a homecoming for Chris, even though she hasn't lived in that country for thirty years.

We got on the water early on a sparkling if chilly morning of a blazing blue-sky day, paddling in amazement once more over that shockingly clear blue-green water. For the first few miles we retraced our initial Temagami trip, hoping to get the campsite of the reflected stars, but it was occupied. The topo map showed no cottages at all on the next lake over, and by lunchtime we'd found a campsite there that we liked even better, on an unnamed island. We settled in.

We named it Perfect Island. (I won't say where, but it's in what was once the family hunting territory of the Egwunas, the Ojibwa family that Archie Belaney first married into.) The campsite was draped across tiers of granite, allowing us to cook on one level, eat on another, sleep on a third. It was fully forested with white and red pines, mountain ash in full orange berry, blueberries, and enough other shrubbery to look as if it had been planted, an informal garden. The island itself was a low north-south ridge, with sloping granite shelves on both east and west sides, giving us *sol y sombra*, morning and afternoon swimming spots and sun perches, observation posts for both sunsets and dawns. At the lake's far northern end, just within binocular range, we spotted a single houseboat; another would putt-putt through on the second day, and a couple of fishermen's outboards droned by, but we would see no other canoes, and no other people, while we were there.

That first afternoon was when I remembered my aunt and diving into champagne. Temagami water is potable (although the authorities, always covering official backsides, are nervous about such claims). While paddling, standard practice is to keep a cup by your seat and dip up a drink whenever you want one. You can drink what you dive into, on the way down or on the way up again. In your mouth the water is not *quite* fizzy with carbonation, but your naked body will swear that your mouth is not telling the truth.

It is the water, I am sure, that Gwen was always looking for and never quite found. This time, late in the season, it was too chilly for extended lolling, but we were in it every forty-five minutes or so anyway. And a full moon in a crystal sky, loon music all night long, the Big Dipper framed in the door of the tent—sometimes you get it right on a trip to the woods. I tried but could not think of a single thing I'd change.

So we sat; bright days, puffy little summer-sky cumulus, an actual tailwind when we went for a tour of our perfect lake. Mostly we sat still and looked, observing the succession of atmospheres, the progression of angles of the light, the rotation of the earth through the long days. "This place fills a reservoir for me," Chris said as we set out on another morning's paddle. "I can run on it for a while, but then I have to come get it refilled."

Thoreau's Journal, July 11, 1851: "The sound of this gurgling water running thus by night as by day falls on all my dashers, fills all my buckets, overflows my float boards, turns all the machinery of my nature, makes me a flume, a sluiceway to the springs of nature. Thus I am washed; thus I drink and quench my thirst."

We did get one gray afternoon, and sufficient rain the last evening on the island to drive us under the tarp for a voluptuously cozy dinner. Then real rain during the night, and fog and mist and early morning stillness closed in for the paddle out. The plan was to avoid a long haul on the last day, so we headed for a campsite closer to our take-out point, cruising back north on the soft gray morning. The haze seemed actually to improve perspective, islands and headlands and hillsides stacked in series in the distance, the progressively lighter shades of blue making depth perception seem more accurate than on a clear-sky day. In moist, soft air, balsam perfume mixed with tinges of birch and pine, sweet and rank at the same time, and every once in a while there was a whiff of what can only be described as pure lake: wetness. (How stupid that we marvel at animals' ability to smell "odorless" water.) We agreed aloud to slow the pace, and mostly did, but found from time to time that we had picked it up again, unconsciously, just from the pleasure of driving the canoe.

Three or four miles of our exit route—about an hour's easy paddling—took us alongside a steep rock face that dropped off sharply into the blue-green depths below. The water surface was perfectly still—no wind, a lee shore—and the rock face a riot of color, reds and grays and yellows and blues, vegetation hanging from the cracks in small communities, the twisted red wood of cedar trunks, sheep's-back lichens in gray and red, mountain ash with vivid green leaves and orange-red berries. All of that glory was duplicated on the surface of the blue-green water, reflected upside down to draw your eye into the depths, the color above duplicated perfectly below but tinted just slightly blue. Down in the depths you could make out the crumbled blocks that had been frost-wedged out of the face above and dropped off the edge, and ghostly white sunken tree trunks making caverns where you knew must lurk the mother of all pike. One six-inch red pine was growing out of absolutely nothing, somehow rooted in a tiny crack in the rock, making us wonder how many predecessors it must have had: trees that almost made it, got started and reached a certain size, and then were simply spit off the wall. Tossed into the deep blue, drinkable, freshwater deeps.

I know a place in the Caribbean called Playa Funchi where black volcanic cliffs make headlands on both sides of a white sand beach and a cove of clearest blue water. You can see a surprising distance underwater with a mask, but abovewater you are frustrated, reflection making the depths unavailable. Unless, that is, you climb those cliffs; then you can see into the water, every shade of blue in the world, parrotfish and squid and grouper and the glittering rainbow of reef life. What makes water so mysterious may simply be this capacity to reflect: to go opaque on you, slamming the visual door in your face, when a moment before, when the light was at only a slightly different angle, all was revealed, in limpid, dreamy clarity. Light and water, flickering on and off, teasing, like memory.

An hour's paddle before our final take-out point, just as the sun came out again, we found a fine small campsite on a rock point, with good swimming. As a visual setting it was actually a prettier spot than Perfect Island, but there was a scattering of cottages around the perimeter of our vision, and yahoos in speedboats. We ate lunch in the sun on the point, with a narrow channel in front of us, two islands to the right, a big, open bay wrapping out of sight around the point to our left. We'd sit there watching as peace and quiet descended, the birds settled in, the water calmed itself and began playing intricate games with the light it reflected onto the opposite shore. Then a speedboat would come by, usually a male adolescent at the throttle, towing another on an inner tube, and all would be turmoil again, and something very like urbanness. Young masters of the universe, out for a ride, just because riding along at high speed in a speedboat is fun. Simple fun, innocent as the day is long and the gas holds out. And then the speedboat would be gone for a while, and peace would settle in again.

Finally they all went to bed and we did too: a tranquil night, then, and up and out early in very heavy fog, unable to see the opposite shoreline of Shiningwood Bay, which we still had to cross. We wriggled along the shore until we could make out one midpoint island, headed for that, and by the time we got there the fog had lifted so we could see both shores easily. We were out, packed, and on the road by 9:25, headed back down off the Shield with just the slightest twinge of sadness. Leaving all that glorious water. Can't wait to go back. Sometimes I think marrying a Canadian was the luckiest thing I ever did.

Footnote, one year later:

PROSPECTORS RUSH TO MINE IN CANADA

> Toronto: In a flashback to the gold-rush era, hundreds of prospectors from across North America raced into a huge tract of Ontario wilderness yesterday, staking claims to mineral rights in an area that had been off-limits for 28 years. Some mining companies hired helicopters to drop prospectors into the bush and bogs. Up for grabs were the rights to dig for gold, copper, zinc, diamonds

and other minerals on a 2,300-square-mile segment of the Temagami region, 250 miles north of Toronto. In June, Ontario's Conservative government decided to allow mining and expanded logging in the environmentally sensitive region after a 23-year freeze imposed by a court because of disputes over Indian land claims. (AP)

15

watersheds

These days the term *watershed* is used more often in the metaphorical sense—a life change, a redirection in the flow of events—than the literal one. We may have lost more in meaning than we've gained in allusion. When we lived closer to the land we thought in terms of watersheds. Most older rural people still do. They know that to understand a place requires knowing where the water goes: where the valleys are, where the streams run, where the natural dividing lines fall. When you live intimately with the land, every creek and river, every ripple in the topography is significant. Whether the slope goes uphill or down determines even such elemental matters as which physical work will be easy and which will be hard. When we urbanize, the watershed is the first thing we lose track of. Interfering with watersheds, as we so frequently and cavalierly do, is one of our more stunning arrogances.

The direction water wants to flow is the fall line; a watershed is a piece of land on which all possible fall lines lead eventually to a single point. Where the water goes is thus a kind of gravity map, delineating the land's true face. Watersheds demonstrate how each segment of land connects with the next, how the pieces interlock. To grasp a watershed is to begin to understand what happens not only with the water but with the nutrients and poisons, the carbohydrates and proteins and minerals, and, therefore, the woods and the wildlife. It is only water and gravity and light, after all—and the dust we came with—that make the natural world. How could anyone not want to understand a creek, a river, a watershed?

When the severity of the northern New Hampshire climate finally began to wear, Chris took on the project of finding us a new place. She did most of her scouting three or four hours to the south. When something looked promising she'd call and I'd drive down and join her for a closer look. Gwen was living with us by then and would sometimes come along, but she was gimpy, facing a second knee replacement, not terribly mobile.

When Chris came across the place we eventually bought, she knew immediately that she'd found a good one, and suggested I bring Gwen with me to see it. Gwen and I drove down and quickly agreed. After a brief tour of the house, Gwen retired quietly to an old overstuffed rocker on the breezeway while Chris and I made a more thorough inspection. It was pouring rain that day, and the breezeway looks out onto the pond. For an hour or so, while we trotted upstairs and down, Gwen rested her aching knees and watched the rain.

That evening, on the way back north, the three of us stopped for dinner, fairly excited, discussing the house. Chris and I tried talking each other out of it, testing reasons not to buy. I voiced concerns about the pond. I'd heard that ponds required a lot of maintenance and could get nasty if you weren't careful. They attracted geese, which could be a nuisance. "I don't know," Gwen said, "sitting there looking out at that pond in the rain, it sure was peaceful." The discussion was over.

So now we "own"—bizarre concept—a body of water. We look at it a dozen times a day, not staring in awe—although in some lights we still do that—but checking it, reassuring ourselves. It is a restful opening in the landscape. It draws wildlife like a magnet. You never know when you're going to see a deer along its far bank, a great blue heron, wild turkeys. In some ways it's the best thing about the place, although the wooded hillside behind it runs a close second.

As I write it is autumn, before freeze-up but with weather too fickle and days too short for relaxed canoe trips. It's the best time of year to walk our hillside, and Chris and I have been tracing watersheds. Although we've now lived here for seventeen years, and over that time I've walked one particular loop on that hill more days than not, I don't yet know every rivulet. Each one that enters our property or originates within it must cross another property line somewhere lower down, but I'm not sure I know where. In the process of tracing them we discovered, on someone else's land on the back side of our hill, a lovely hidden beaver pond we had no idea was there—not half a mile from the house—and a tumbling brook that makes Japanese painting after Japanese painting all the way down to the valley below. The next fair day we plan to take a lunch over there and listen to it while we eat.

The stream runs into Creamery Brook, as does all the water from our hillside, although our runoff follows at least three different paths to get there. Creamery Brook runs into the South River, which runs into the Deerfield, which runs into the Connecticut, which runs into Long Island Sound, which is connected to, or part of, all the rest of the water in the sea, in all the seas. I like to think of it as a fairly long chain of molecules—but then if you start thinking that way you can connect everything in the universe to everything else, can't you?

Someday I'll walk from the height of land behind our house to the Deerfield, following water. It's not that far, no more than ten miles. Maybe I'll count tributaries. God knows how many feeder creeks there are in that distance, coming in from either side. On my wall is a map of the United States that shows only state lines, lakes, and watercourses. America looks like it has varicose veins. We are rich with water, if we can figure out how to take care of it. Or if we can figure out how to leave it alone.

When I began competing at swimming again, in my late forties, I quickly grew to love training over long distances in workouts and racing long distances in meets, loved it all too much, trained at it too hard. After a few years, accumulated wear and tear began to take the fun out of it, and I retired as a racer. In the meantime I'd discovered free diving in Caribbean waters, with no equipment but mask, snorkel, and fins. That's a way to swim I'd somehow missed, and is what I was born to do. I'm not generally a saltwater enthusiast, but playing along a reef in thirty feet of clear water, surrounded by the soup of life that a reef always draws, is the most enjoyable swimming I've ever done.

It surprises me now how many strange swims I can remember. I've interrupted more than one long motor trip to explore some piece of water, requiring extraordinary patience from my traveling companions while I changed (or just stripped), went for a swim, and came back to the car to add wetness to the other little annoyances of automobile travel. (If you are driving through woods or mountains and see parked cars beside the road, and it isn't a trailhead, there will almost always be a nice piece of water nearby. You can depend on it.) There have been outlaw excursions, moonlit sneak-ins, trespasses into forbidden waters, and more stinking, opaque, piss-temp motel pools than I like to think about. I've gone into some really ugly creeks. There are people a lot crazier about swimming than I ever was. I have friends who've swum around Manhattan, for which cholera shots are advised. They say it's not so bad.

When I think of the generic act of swimming, what always pops into my head is diving off the lifeguard tower, more than forty years ago. I'd bake in the Texas sun until I felt as if I might faint from the heat, then dive in and hold a long glide through the enveloping coolness, imagining if not actually feeling a sizzle from my skin. I still love to do that, to plop into a quiet lake and glide as far as I can, then stroke along in a leisurely crawl until I'm all tired out. That's still the ultimate tranquilizer. Now my shoulders begin to ache within a hundred yards, and if I keep it up too long the next day's paddle will not be pleasant. Sometimes as I'm falling asleep I get a sensory flash of swimming in still water, bright sun flashing on my face as I turn my head to breathe, and it fills me with a wave of longing, missing that, missing those lovely distances. Paddling them, however, is in some ways better, and I can still free dive. Lose some, win some.

I trained at swimming rigorously, according to the established principles of athletic physiology, of which I grew very fond. Once, though, I went to my evening workout in a black depression, generated by some spiraling series of financial or professional setbacks the details of which, unlike the swim that followed, I no longer remember. The coach had posted a detailed workout, but I wouldn't even consider it. "No thanks, I just want to swim," I said, and went off to a gutter lane to be left alone.

I started cruising slow laps, paying little attention to what I was doing, brooding instead about the unfairness of things. The water washing across my face felt somewhat like embarrassingly appropriate tears. After a sufficient number of self-pitying laps, I began working the water more aggressively—unthinkingly pouring my resentment into it, I suppose. It rewarded me by allowing me to move through it more swiftly. My pace inched up to workout levels. I began watching the pace clock, realizing that I was posting quite respectable times. Something mysterious was raising the upper level of possibility. Perhaps the water finally drew itself sufficiently to my attention, and offered enough challenge—and enough reward—to interrupt my interior harangue at the inequities of life. For an hour or so it gave me a nice vacation from all that.

Before long I found myself grinning each time I swirled into the flip-turn at the end of the lap, ducked under, and blasted off in the other direction: another unit of effort bitten off, energy deliciously spent. I ended up swimming farther and harder than a normal workout, and drove home in a calm, quiet ocean of peace.

Water has consistently done that for me, by means I've never clearly understood. When I was a child it was an entirely other world, a different place, that I needed, or thought I needed, badly. I had the luck to discover it early, and found it manageable in ways that the dry-land world was not. I had a small gift for making use of it. That gave me an advantage while I was in it, one I wasn't about to give up.

The thing about the ritual morning plunge, the entry into water that provides that small existential moment, is its total privacy. Swimming is between me and the water, nothing else. The moment the water encloses me, I am, gratefully, alone.

Trying to pull to the surface a mental image of the Jerome family canoe, I am startled by another wisp of memory. My stepfather Ott never betrayed interest in any recreation other than woodworking, except perhaps for the Indianapolis 500, which he had once attended. Tucked away in the rafters of his woodworking shop, however, was a pair of gigantic canoe paddles. He never explained them—we did not talk a lot—but I eventually learned that he'd grown up in Illinois, on the banks of the Mississippi, and with his teenage peers had spent a great deal of time in canoes. He was a tall man and quite thin, but he had massive shoulders and arms, Popeye forearms and ham-like hands. The oversize paddles, which he'd carved for himself, implied the source of that peculiar physique. They also implied something about his personality, his nonworking life, that neither Jud nor I ever dreamed existed. He never mentioned canoeing in my hearing, never sold it to us the way Ralph had. Why had he given up that enjoyable pursuit? Why did he keep it a secret? Or was it only that I never asked the right question?

My legacy from him was an indirect one. He was "handy," made things work, could fix things. I spent considerable time with him holding tools and otherwise being required to help. As a result I acquired, in addition to a certain manual facility of my own, an almost religious respect for clear mechanical solutions. When something works out mechanically I get a moment of smug enjoyment, followed by a wash of pure relief: see, reality works.

Chris, on the other hand, has a far better grasp than I of the biological aspects of the natural world. I think it comes from her powers of empathy. She has the capacity to place her imagination inside the bird box and know what the nestlings are up to, what they must be needing at the time. I only get a glimpse. In late fall, watching the rain slowly strip the foliage off a tree at the end of the pond, I try to acquire her viewpoint. What would it be like at the leaf? There would be some kind of comprehensible process of physical detachment: a gradual increase in weight, a wetting down and softening of fibers, and then their tearing away under the actual hammering of rain drops. The farther out from the stem the drop hits, the greater its leverage on the leaf. There you go, physics again, solid joints and molecular explanations, as concrete and comprehensible as the lever by which, if you would just give him a proper fulcrum, Archimedes promised to move the world. That's what has always made me a sucker for the physical: that through its mechanical clarifications one might, finally, understand everything. Or something.

• • •

I am not too successful at absorbing the natural world. My attention fades. (Nature always fails you, as Thoreau points out.) When I take my retinas to the woods, my hope is that they're getting the good of it no matter what other trash is occupying my consciousness: raw data over interpretation every time. It occurs to me that turning conscious attention onto it may be counterproductive, that, as with concentration, to think about it is to lose it. Maybe trying to grasp it—this mysterious thing that is pouring out of that world and onto my fortunate nerve endings—is also pushing the river. Perhaps what our saints are trying to teach us is to quit caring so much what our minds are occupying themselves with—that too being a kind of river. To be conscious is to feed the monkey; the monkey is going to be fed, stop worrying about it. How do you do that?

Walking in the woods at dusk on a windless evening, I see no deer but keep imagining I do. It's almost impossible not to imagine them, drifting through this glade or that, moseying but alert. Down in the valley I hear truck traffic and chain saws—our neighbor seems to be tilling his garden—but up here, within all that, is perfect quiet, the hillside a cone of silence within a valley of noise. I imagine myself drifting along the path as quietly as a deer. I can't, of course, can't be that quiet, or that alert. I wish I could. I wish I could be in these woods the way a deer is in these woods.

What I think I want is just to be here, enjoying the woods' own indifference. The vast, shuddering, thunderingly silent indifference. I would fall on my knees and give thanks for that: thank you for your grand, loving indifference. I try to keep these practices from going symbolic on me, and find myself thinking about that, and so lose the thing itself; and fail again. The Japanese tea ceremony, I think, is about preparing a cup of tea. I suspect that the more acutely you can keep it about preparing a cup of tea, the more volumes it will speak, the more its significances will reverberate. That may be the point—to make the significances reverberate—but while it's happening you may lose out on having your cup of tea.

Chris says she can reproduce in memory almost nothing of her daily life but can recall in some detail almost any day she's spent in the woods.

One carry on our long Adirondack trip began at the crumbling wreckage of a dock at the end of Lake Utowana. It once served the *Tuscarora*, a steam packet that collected passengers bound for W. W. Durant's country club on the next lake up the line. They'd come from Raquette Lake by rail—only half a mile, on what was reputed to be the world's shortest standard-gauge railway, built by Durant in 1900 and abandoned twenty-nine years later. The portage follows the old rail line.

The abandoned steamboat landing, once the site of a great deal of human activity, is now in de facto wilderness, a curious sight in its present context: a few sunken timbers, twisted bolts, a hint of structure. We arrived at it on a sunny, humid day, the air close, cicadas and other insects buzzing. Homer's mosquito hum and wavering dragonflies, as imagined by Thoreau, came to mind. The days—the sun, the humidity—had to have been the same when the landing was in use ninety years before. I found it oddly moving to think about that, the sheer atmospherics. The day you're getting in the woods is exactly like some day in that place—wherever the place—a hundred years ago, a thousand years ago. Museums and restorations and re-creations can't give you that. What makes the wild places so radiant is that they are simultaneously ancient and fresh. They are renewing themselves—creating themselves—every moment: they *are* creation, creation itself, in progress before our very eyes.

I've mentioned here more than once some pleasurable small moment when circumstance makes me conscious of the planet under me, and of where I am on it. It happens more often in the woods or on water than anywhere else. You can't think of where you are on the planet without thinking of the planet in motion. That puts you in real time. One thinks of going to the woods, the timeless woods, as a way of escaping time for a while, but it's really the opposite. Going into the woods stops everything *but* time, everything but the world. It isn't the world we want to stop, it's all this scurrying we do on its surface, that doesn't get us any*where* or, really, show us any place.

Scurrying through the nonworld. Stop the world, I want to get on.

16

blue rooms

Man is an obligate aerobe.

—HIPPOCRATES

The hotel's scuba-diving operation moors its four big dive boats in a line well out from the end of the dock. Getting all four stowed away for the night is a complex ritual. One dive master waits on his boat while the other crews moor theirs, then goes down the line, fetching them and delivering them to the dock. He then takes his own boat back to its mooring. He leaves the motor idling while he scampers forward with a boat hook to snatch up the floating line, then darts back to cut the throttle. All systems secure for the night, he dives in and swims back to the dock. It is virtually the only time I've ever seen a dive master swimming on the surface, without scuba gear.

Chris and I would enjoy this ritual every evening from our second-floor balcony. The dive masters take turns with the mooring duty, and come in all shapes and colors, but all are dark from the sun. One memorable dusk we watched a particularly tall and lithe young man perform the mooring ritual with unusual grace. His boat in place, he dived in and swam slowly back to the dock, then hauled himself out and, standing on the end of the dock in silhouette, backlit by the evening sky, rinsed the salt from his body with a freshwater hose. He was wearing only a tiny wisp of a swimsuit and spent a long time at his ablutions, sluicing himself from head to foot over and over again, obviously enjoying it.

So did we. He must have been six and a half feet tall, with a willowy swimmer's build: remarkably sexy, actually, a veritable Greek God, Caribbean version. He would play the hose on his head, under his arms, inside the front and back of his vestigial swimsuit, seemingly oblivious to the possibility that he was being watched. The hose was turned low, a bubbling clear stream rather than a spray, and its outpourings outlined his dark body with a glistening, shimmering sheen, which he squeegeed off with one hand while playing the hose on himself with the other. He couldn't seem to bring himself to quit, and we didn't want him to. It was one of the most erotic performances either of us had ever seen.

We were in the Caribbean because a beneficent fate had handed us a somewhat unexpected winter vacation. I'd just finished a project that not only occupied an indoor winter but also left us with a little cash on hand. The end of it came in a long, high-intensity grind, in the midst of which I observed to Chris that we would soon be extremely ripe for a break. The following wide-ranging discussion ensued: "Let's go to Bonaire," said Chris; "Okay," said I. What the hell, it felt like playing with house money.

Bonaire is, or used to be, our Caribbean island of choice. In one period of relative prosperity we managed to go there five springtimes in a row, but hadn't been back in six years. The island is one of the group, with Aruba and Curaçao, known to travel agents as the ABCs. The other, larger two are cruise ship islands and therefore damned; on our last previous visit Bonaire hadn't yet achieved that dubious goal. This was not surprising. It is your basic desert island, featuring mostly cacti and other thorny fauna, volcanic rock, lizards, and wild goats. I once showed snapshots of its interior to a friend, who assumed they were of west Texas. Devil's Island, the infamous French penal colony that once stood as a metaphor for tropical hell—see *Papillon*—is just down the coast.

Bonaire also has prodigious bird life, and a leeward shore that drops off within thirty yards into a steep-sided reef leading down into blue depths of what is arguably the clearest water in the Caribbean. In that water is fish life to surpass its birds. Its entire underwater coastline, to a depth of sixty meters, is protected national park. We have a Dutch friend, a birder and scuba diver, who once asked a travel agent about Bonaire. "Well," said the agent, "unless you speak Dutch and like to dive, and maybe look at birds, there's not a damned thing to do there." Speaking Dutch—or Spanish, or Papiamentu, the other local tongues—is not required, but otherwise the travel agent was correct.

The mental health establishment should study the therapeutic effect of opening an airplane door in winter in the Caribbean: the first faceful of sultry breeze, a blessed whiff of the sensual assault to come; itchy dry winter skin starts healing instantly; sweat pops, and is welcome. From that point forward all sounds, smells, colors seem suddenly richer, hotter, more intense, at least to muffled northern senses. The Caribbean starts peeling your clothes off.

We got to our hotel in time for a sundown swim and a rum punch before dinner, not necessarily in that order. The air no longer hurt, as it had in the New England we'd left that morning. Darkness, when it arrived, was plum colored. The thin sliver of moon was so situated at that latitude that its two upright points were exactly level, horizontal, an attitude I couldn't recall ever seeing before. Orion, Chris pointed out, was perfectly framed in the arch over our hotel room balcony, where we sat, bare feet on the railing, watching the lights of distant boats appear and disappear over the horizon. Certain psychic sphincters—the ones that had begun to close as we left Temagami seven months before—began to reopen. In the soft tropic night I began to feel as if I were composed entirely of skin.

But then the kind of swimming we go to the Caribbean to do is sometimes called "skin diving." I like it particularly because of that, because it not only gives you the feeling of free flight in an enormous blue room but is also a kind of naked slithering through limpid and entirely comfortable waters. And then as you go deeper it becomes a kind of hugging machine, a great, welcoming embrace from the physical world.

Some people find swimming claustrophobic, and I understand that, but for me it is the absolute antithesis of, the treatment for, claustrophobia. Scuba diving, however, is not; in fact I have felt claustrophobia's clammy panic while wearing scuba equipment, not from the mask or the water but from the enveloping, ill-balanced weight of the gear. Scuba diving is entirely too passive, an experience handed to you by hardware. What it gives you is access to the visual world of underwater while at the same time clamping you off from the sensual world, the proprioceptive world. To conserve air you must conserve energy, which means you move as little as possible. You hardly get to work the water at all. Scuba diving is making love in long underwear.

The other joy, true of both skin and scuba diving, is to put oneself in the midst of fish. Sometimes hundreds of them will be swirling around within six feet of your face. They stare at you, you stare at them. They stay just out of reach, darting away long before your hand gets near, if you are silly (i.e., human) enough to try to touch them. You can't "touch" a fish, of course, but in fact you're already touching them, generating currents, sending signals that they feel, whether you can observe their response or not. I can't help dreaming that if one spent enough time with them, perhaps you could learn to feel those signals too, coming back the other way. Through pressure, through subtleties of current, you might begin to feel touched by the fish. You might then begin to understand the web between you, the contact established at some subtle level with all those wild creatures the moment you go into the sea.

Bonaire was "discovered" in 1499 by Amerigo Vespucci, although Arawak pictographs in local caves imply that the island didn't exactly spring into existence with the arrival of Europeans. It was colonized by the Spanish in 1527 and became a Dutch colony in 1636. Attempts were made at agriculture, but virtually the sole significant export of the island has always been salt, and slaves were imported early to work the salt pans. British and French privateers harassed the island during the Napoleonic Wars; the British occupied it briefly but seem to have been baffled by the place, and leased it to a New York merchant for $2,400 a year, three hundred slaves included. In 1816 the Dutch took it over again, abolished slavery in 1863, then parceled out the island and sold it off. Slave huts still stand, tiny stone boxes with doorways four feet high, now preserved as photo opportunities for tourists. It is unimaginable that the huts might have been habitable under the tropical sun, but then slavery has always kept long hours, and occupancy during daylight was probably not required.

After slavery Bonaire snoozed, a subsistence island, until the 1920s, when Curaçao and Aruba built oil refineries. Bonaireans began migrating to work in that industry, an era local history refers to as the "money-order economy," for the wages that adult males sent home. When automation wiped out the oil jobs, Bonaireans came home, essentially unemployed again until tourism came along.

There was an American naval air base on the island during World War II and an internment camp for Nazi sympathizers, which happened to be located on the site of our favorite hotel, the Flamingo Beach. The oldest hotel on the island, the Flamingo Beach has always had a familiarly compoundish air to it, a shady enclave in the otherwise harsh landscape. Internees were British, French, and Dutch, probably plantation class, probably not poor. The compound's western boundary is the sea, where the sun set then too, every evening, in the usual blaze of tropical glory. Rum punch is not a recent discovery. Internment there may not have been all that cruel. I see rumpled linen suits and long cigarette holders, but then I've always had a weak spot for bad novels.

In 1954 the Netherlands Antilles, of which Bonaire is a member, achieved autonomy (some islands later gave it back), which brought in a trickle of investment. The salt industry was modernized and automated—it is now solar powered and alleged to be environmentally sound—and the airport upgraded in pursuit of tourism. The Bonaire Petroleum Corporation built a transshipping terminal on the leeward shore for the loading of oil tankers (done with great care to avoid damage to the reef, and therefore to the tourist industry). In the 1960s two huge shortwave radio broadcasting operations moved to the island. Employment seems to be fairly full, the island almost bustling, at least by Caribbean standards. Condominium development had taken off since we were there last, and downtown Kralendijk spruced up. One cruise ship did dock during our stay, but its passengers couldn't seem to find much to do.

Our previous visits had always been in early March; this time we didn't get there until April, and the angle of the sun had advanced far enough northward that the uninhabited islet of Klein Bonaire, just west of the main island, interrupted the sunsets, so we couldn't quite see the point where the sun finally winked out beneath the flat line of the sea. That was frustrating, wiping out the possibility of seeing the green flash. A month later also meant a month hotter, resurrecting strategies from my Texas childhood summers, where shade was almost as precious as water. Anything dark is going to be painfully hotter than anything of lighter color. If you are barefoot—and you usually are—the red sidewalk tiles will be more painful than the yellow sidewalk tiles. In April, we discovered, there is more than enough light.

After a couple of days we gave the diving and our aching bodies a rest and explored the windward side of the island, unswimmable most of the year. ("Please, No Sturdiness" warns one brochure.) It features, as the leeward side does not, the now-standard wrack line of floating plastics: wads of Styrofoam, thong sandals and abandoned Nikes, seat cushions, and thousands upon thousands of package-goods bottles, an accumulation more frightening than the thundering surf and mysterious undertows. In our lifetime the romantic pastime of beachcombing has gone from a way of adding interest to a stroll on the beach to a visit to a garbage dump, simply because of plastic's insistence on floating.

I picked up two containers, before losing heart, and read their labels. One was a German toilet cleanser, contents toxic as hell; the other was an American feminine deodorant, to be sprayed from a distance of six inches on the surrounding genital area but by no means allowed to contact any moist areas. I wondered how the bottle's contents smelled or tasted to the fish, when it was cast adrift and the remaining effluvium began to disperse into the surrounding seawater. I wondered about the professional satisfactions and rewards for the people who conceived this product and invested untold (but billable) hours in determining just the right gentility of expression for writing the directions for its use—or for convincing a nervous populace that such a product would fend off social disaster.

Having worked for a time in that business, I understand thoroughly the level of dissociation with which one performs those labors. I learned without too much trouble to achieve it myself, on demand. Dissociation is surely the key. We are born into the natural world and then choose, or are taught, to dissociate from it. Those who do not dissociate are our savants. (Usually they lead uncomfortable lives, and few of us would choose to emulate them.) My own headlong attempts to achieve that dissociation have been more successful than I can bear to think—for a while there I wrote ads selling

Corvettes, for Christ's sweet sake—but I've been lucky. For some reason the phenomenal world, outside the sweaty little fluorescent-lighted cubicles where the dissociation goes on, has always had the effect of forcing reassociation. I look forward to spending the remainder of my days in further pursuit of that, as long as my nerve endings hold out.

Those nerve endings are not necessarily distributed equally, which always comes as a shock. Chris and I, for example, have markedly different olfactory capacities. Each of us gets information denied to the other, and when we do get the same information we sometimes interpret it in different ways. There are flowers Chris loves that I can't smell at all. She likes to have a bowl of narcissus on the windowsill for their perfume; when they bloom I come running downstairs looking for an electrical fire.

Also, nerve endings come and go. My hearing is still technically okay, but that doesn't mean I hear well. Perhaps I just don't listen, and nerve endings have nothing to do with it. I am handicapped in the matter of taste buds, delicate flavors being wasted on me. My visual acuity varies except at short range, which is now mostly gone. Vision would be king of the senses for me too, as it is for most people, if it were not for touch. I get things best through my fingertips, the palms of my hands, the skin of my face. Almost all of my attraction to swimming seems to me to have grown out of that, another way of touching things with my skin. I would not intrude on your space for a moment, but I am truly comfortable only with people I feel free to touch.

Sensory day, Bonairean version: breakfast on the terrace (rich Dutch coffee), bananaquits and ruddy turnstones fluttering among the tables, fishing boats putting out to sea in morning light so soft you can bathe your face in it. A slow drive north up the leeward shoreline, volcanic cliffs in deep shadow on the right, achingly bright sea—turquoise shoals, white sand shining through blue clarity—on the left. Cut inland over cactus-covered hills, in midmorning stillness and growing heat, to Washington-Slagbaai National Park. Stop off to get your blood oxygenated at Chikitu, a windward inlet with pounding surf, where salt spray opaques your sunglasses in twenty seconds. Make the slow, dusty, bumpy drive around the north end of the island, keeping an eye out for the parrots that flash green in the underbrush. Seek welcome shade, briefly, at one of the inland waterholes, where you may see sea eagles, troupials, parrots, and parakeets, along with the more familiar faces of doves and mockingbirds. Get yourself uncomfortably hot, dusty, sweaty, because swimming comes next.

Pick your spot on the leeward side for that: our old favorite Playa Funchi, for example, a deep cove with a minuscule rocky beach, black cliffs along both sides, backed by a flamingo-splotched inland pond. Grab mask and fins, wriggle out through the shoals, wind your way among huge coral heads to the small water-line caves on either side, or simply cruise over the white sand shelf between, in twenty to thirty feet of water, before the usual wall drops off into the usual huge blue room. Frolic in the enormous space of that great blue nothingness; explore personal limits of depth and of breath holding. (Breathe deep; breathe, in fact, a great deal.) Enjoy, at depth, the weight of the atmosphere pressing on your body. Salute barracudas, squid, yellowtails, sergeant majors, needlefish, various jacks, trunk-, angel-, file-, parrot-, and trumpetfish.

Get out, sun, go back in, get out, picnic in the shade—defending lunch against extremely friendly iguanas—and go back in. Assess level of waterlogging; make the dusty drive home in hammering heat to cool room and cool shower (the skin craving even more water) and nap, and give the rest of the hot bright afternoon over to relaxations of one skin-related sort or another. Consider, over cocktails, the amazing day, the amazing light, the amazing exposure of the integument to a benign but spirited and unequivocally intense physical world. Welcome the dusk—however brief—when the earth creeps deliciously, as Nicholson Baker points out, into its own shade.

Even a week of Caribbean free diving brings physiological change: I can feel myself rounding into shape for it, and exult in that. There is always a turnaround day, before which I am having to work a little too hard and am not quite comfortable at it, after which it is natural and easy. Once that happens, there's always another fifteen or twenty seconds available down there, and I begin going deeper, experimenting with how deep I can go and how long I can stay down—which, never mind looking at fish, is for me the secret thrill at the bottom of the whole enterprise.

Free, breath-holding diving is actually a competitive sport, although very loosely organized. Its participants slide down a cable, hanging onto a weighted sled that drags them down as quickly as possible with a minimum expenditure of energy. They mark their deepest point and then swim to the surface. The last time I looked the record was 341 feet, held by a woman. I once proposed, for a magazine piece, to set the breath-holding depth record for a sixty-year-old man. I planned to do it as pure spoof: there was no such record, so anything I did would qualify. No fancy cables and sleds, I'd just slip over the side of a boat with a concrete block in my hands, hang on until the pressure became uncomfortable, let go, and swim back up to the surface. Piece of cake.

The project inflamed my imagination for a few days. I would contact the Guinness people, find out what was required to legitimize the effort, rig up the appropriate ropes and measuring devices. I'd invite friends along to help with the logistics, perhaps even get some magazine to pay their expenses. We'd have a hilarious free Caribbean vacation, set a record of forty or fifty feet, and make fun of everybody. Then I looked in Guinness and saw the asterisks, indicating accidents and fatalities. Breath-holding diving, according to the authorities, is an extremely ill-advised endeavor. Funny, but when I'm doing it I always think that if I lived in the islands and did it every day I would live forever.

In a recurring dream I am swimming underwater, enjoying it but running out of air, and just at the point of desperation a kind of toggle switch is somehow thrown and I can breathe the water. I can stay down there. As a kid it was always a major disappointment not to be able to stay longer, where it was so nice and cool and different. This should have made me a scuba diver but did not.

I suppose that's what *The Water-Babies* is about, that essential Freudian fantasy, very like the dream of flying—which surely comes from the same source. If I'd read that book as a kid I'd have been sucked deeply in until the hero's moment of transition, when he acquired the ability to breathe underwater. That would've been all I was interested in, the technical part of the story—which of course is left out. After that I'd immediately have been driven right back out of the book by its preachy Victorian moralisms. It's worse than the Boy Scouts for that.

Deep breath-holding diving is dangerous because highly technical but not clearly understood things can happen to your physiology at those human limits. "Generally speaking, the 100-meter mark defines the bottom of the euphotic ('well-lit') zone, the most productive part of the sea," says James Hamilton-Paterson in *The Great Deep*. "Coincidentally, it also marks the theoretical limit to unaided diving by creatures of the upper air. Certain free divers have exceeded this by a few meters . . . but it is at around this point that pressure collapses the lungs. Well before that, the strain of keeping them inflated will have begun to rupture small blood vessels." I've never gone particularly deep—thirty or forty feet is surely my record, which is nothing to any spear fisherman or pearl diver—but I've had glimpses. I've gotten drunk enough on the experience—an early stage of nitrogen narcosis, probably—to begin getting the weird sensation I might not have to come up at all.

Of course you can drown in a teacup. For us air breathers water is a go/no-go medium: above it, life; below it, not-life. Going into water, using it for one's own purposes, is a kind of thumb in death's eye. Continued reimmersion brings familiarity, confidence, if not contempt then at least the illusion of mastery. And intimacy: making it not so strange. Ah yes, you are always thinking with some tiny part of the back of your mind, that could happen to me, here, in a moment. Probably won't though. Once you're comfortable at it, the water you have all about you becomes somehow more important, or at least more enjoyable, than the air you are rapidly running out of. The discomfort of the respiratory drive doesn't quite get bad enough to bring you to the surface. If it hurt that bad you'd quit, you'd get out. You keep going back. You quit only when you get chilled.

What then becomes most rewarding to do is just to hang there, the need for air building and building but inconsequential somehow, a deerfly buzzing around your head. You see something else interesting—an eel? a flounder? what is it?—and kick on down another ten feet, peering more closely, intrigued, and, simultaneously, amused at yourself for continuing to linger. There is a curious seduction in that, in cavalierly dismissing the very possibility of actually running out of time. This has to stop in a moment, doesn't it? You do have to go back up. In a moment, in a moment. Rapture of the deep, they call it. There are few enough raptures still available. We seem to evolve, or mature, out of them.

And then comes the almost casual return to the surface—always ecstatic but always slightly sad—and the explosive release of dead air, and the sweet inhale that announces the next cycle: now you can go back down there again, into the hugging machine, the big squeeze, the total, whole-body, ultimate firing of the pressure receptors, the receptors of touch, the nerve endings of acceptance. You go again, and again. Come up to the surface, get a breath of that sweet air, go back down again. All we are are nerve endings. All we get are days.

epilogue: the river within the river

Sound travels underwater better than it does on the surface, and fish hear extremely well, at least as far as we can know such things. They may not have ears to hang eyeglasses on, but they do have acoustical equipment, including a lateral line of pressure sensors running down the body from head to tail. These hairlike sensors, sensitive to currents and changes of current, surely also pick up the pressure vibrations that we read as sound.

This information about fish anatomy comes from James Hamilton-Paterson, a wonderful writer who for a third of a year at a time has taken his living out of the sea by skin diving, and has therefore thought a lot, clearly, about sensory information. He's particularly keen about sound, rejecting scuba apparatus specifically because its bubbling and sucking noises keep him from hearing what he wants to hear. He describes hearing whale song while free diving: "It is more than just the nakedness of wearing next to nothing, and it is more than vulnerability. It is the sensation of animal messages passing *through* one as if, being seven tenths water, one's body were transparent."

But then hearing, particularly underwater, is a curious business. "A bat 'sees' with its ears with great precision and at speed," Hamilton-Paterson points out. "In short, there *is* no such thing as darkness." If there's no such thing as darkness—sensors coming in all shapes and structures, capable of reading signals unimaginable to our human limitations—then at some level there's no such thing as silence, either. I find this a depressing line of thought, and suspect Hamilton-Paterson would too.

With such an array of sensory gadgetry, a fish must easily be able to locate the river within the river, the region of fastest flow. To zone in on it precisely would perhaps be to reduce it, theoretically anyway, to a single point, a kind of one-dimensional line winding through the water. I try to imagine how its rushing current might feel—sound? look?—to a fish. If its swirls and meanders come to the fish as sound, that would make it the melody, wouldn't it, buried within the harmonies of the river?

In the matter of hair and hairlike structures, it might be noted that for important meets, competitive swimmers shave their limbs and bodies to reduce drag. The performance gain is estimated at 1 percent, a highly unscientific figure; the larger gain is probably psychological. The first time you dive in after shaving down you feel as if you might squirt right out of the pool again. You feel as if you are within the water in a whole new way, almost bonded to it—and cutting through it faster than you've ever done before.

Maybe what I'm driving at is just that water is our best sensory fit. God knows it is beautiful enough to look at, in almost any setting. It is allegedly tasteless and odorless, but only in a laboratory setting: certainly nothing tastes better when you're thirsty, and animals can smell it for miles—or smell its setting, smell the fecundity and organic richness it unavoidably generates. Japanese gardeners are famous for "tuning" their fountains and the riffles in their streams for optimum sound. Nothing is more soothing than the sound of surf.

Then there's proprioception—call it "feel," as opposed to "touch"—which tells you what position your body is in and what it is doing, particularly when it is in motion. Water, with its constant pressure on the entire body and its gentle resistance to movement, is always telling you exactly where you are; it *locates* you, sensorily; its resistance informs you exactly what your limbs, your musculature, your physical self is up to. To be afoot on dry land, by contrast, is to be dependent on the sensors of the soles of the feet—and the air itself—to give you any sense of locus. Vertigo is just around the corner.

For swimmers, though, the sense that counts is definitely touch: water as licit pleasure, a very large and direct physical sensation that is not only legal but is even socially acceptable. Not orgasmic, at least to anyone I ever heard of, but definitely erotic. Marathon swimmers—female ones in particular—are well known for indulging in steamy fantasies during long swims. I have swum in one-hour races myself, and while the fantasies that engaged my attention during those painful and painfully boring exercises had no directly identifiable sexual component, I definitely did have fantasies. In one of them, three-quarters of an hour into the grind, I began imagining the indoor pool I was swimming in was surrounded by snow-covered peaks—specifically Himalayan ones, for some reason, perhaps having to do with repeated laps and the whirling of prayer wheels.

There is also a subtly erotic pleasure to sliding over the surface of the water in a canoe. Even that, the silky smooth covering of distance, cossets you with sensation. The same is true of board sailing, of course, and waterskiing, sheer speed multiplying the sensations by several orders of magnitude. In the case of waterskiing it comes at the expense of a level of noise that to my tastes doesn't suit the surroundings, but I suppose there is a kind of erotic violence in it, the buzzing motors and all, as there is in speedboat rides.

Clearly we who love water feel we are loved by it: dive into it and it touches you all over, instantly, in your most private places. The ultimate mouth, or womb. It seems to love your skin; your skin certainly loves it. It makes you *sleek*, at least temporarily, however ragged and hoary you otherwise may be. I'm sure that was the first thing I noticed about it as a child. Being taken swimming in the Mountain Fork River in Oklahoma, going for lessons at the municipal pool in Texas, I knew at some level I'd found something I wanted with what at that age amounted to a quivering lust: water on my body; my body in water.

But to tell the truth, that particular aspect of it came to me only now, with the phrase "touches you all over"; when those words formed in my mind I realized instantly that that was what I had loved about it from the beginning. I wanted to be within that medium, wanted it surrounding me and supporting me, squeezing me, squeezing harder the deeper I went. I'd just never quite admitted it to myself. Or I repressed it, perhaps in fear of Gwen's prudishness. What she taught me about water she taught with a wink and a nudge, never an explicit lecture.

Funny how you can put a memory away like that. But then they are written on water, aren't they?

One summer during World War II Gwen's sister Orpha rented a vacation cottage in the Texas hill country, and Gwen and I joined her for a couple of weeks. The cottage was half a mile from the Guadalupe River, and I would walk down to it every day, to a combination dam and low-water bridge that provided the only available swimming hole. Above the dam the water was only about four feet deep, and below it didn't even come to my twelve-year-old knees. There was no place to dive, no rope swing, nothing to play on but a road with a couple of inches of water running across it. Not much you could do with that.

The proper water level was maintained, however, with the help of a culvert at one end, about twenty-four inches in diameter, that carried excess water beneath the dam. It spewed a constant stream on the downstream side, foaming explosively out onto a sheet of slippery granite. There was a good kids' game in trying to wade up against the current, a foot or two deep. You'd eventually get knocked off your feet and swept exhilaratingly downstream for a couple of dozen yards, and then clamber back to try it again. That amused us for a while.

The entrance to the culvert on the upstream side intrigued us more. There was no grating over it, no protective barrier. We were earnestly warned not to get too close, or we might get sucked in and drowned, but the adults who warned us weren't always in attendance. We'd toss sticks into its sucking maw and run to the downstream side to see them get spit out again. I would time their passage: hold my breath, drop a stick, run to the other side, wait until the stick had reappeared before breathing again. Another kid and I talked it over at great length. You could probably do it. If there was any kind of snag within the pipe you'd be dead, of course, but the sticks always came out the other side. There was probably no obstruction. It had to be possible. It would be glorious fun—a swoosh through, ten seconds of darkness, and then ejaculation out into that downstream spew.

We never tried it. We speculated about it for hours. We launched and observed every substitute for our living bodies that we could think of, we argued, we convinced each other it was a cinch. It came to stand as a symbol for all our prepubescent frustrations, all the withheld permissions and postponed gratifications, everything we couldn't have that we wanted so badly. It remained undared. Another year's worth of hormones and I'd probably have given it a shot. I was never very good at withstanding that kind of temptation.

Now, in late adulthood, I dream of waters almost every night. My unconscious invents a piece of geography and fills in at least some of the sensory detail: colors, water quality, angles of light. Sometimes I swim or canoe with great sensual pleasure, but more often than not it's the traditional frustration dream, in which there is something I want to do but can't quite accomplish, some reason I can't use the water the way I want to, the water that's right there. I go for a swim but am unable, for some complex reason or other, to complete a stroke. My arms don't work right—the neuromuscular patterning has been interrupted, closure is denied—or the pool is too shallow, my fingertips dragging the bottom, fouling in mud or weeds.

Quite often in these dreams there is some variation of a waterfall, to be swept over which would obviously be fatal. I nevertheless somehow stop paying attention and get caught in the current, and over I go. After the traditional moment of panic—that quick little realization that this time I really am going to die—I see that there are other swimmers in the falls, and they are enjoying it. They do this all the time, it is just another thrill ride. After a shudder of fear I come out of it all right with the rest, survive, and scramble back to shoot the falls once more. Squirt under the roadway and pop out on the downstream side of the dam, grinning in the hot sunshine, and go scrambling back to the upstream side to shoot through the void again. Alive. Alive.

Eros: my God, I never knew. It *was* a love affair. It still is.